2010
Yearbook of
Astronomy

2010 Yearbook of Astronomy

edited by

Patrick Moore

and

John Mason

MACMILLAN

First published 2009 by Macmillan
an imprint of Pan Macmillan Ltd
Pan Macmillan, 20 New Wharf Road, London N1 9RR
Basingstoke and Oxford
Associated companies throughout the world
www.panmacmillan.com

ISBN 978-0-230-73605-4

9 8 7 6 5 4 3 2 1

A CIP catalogue record for this book is available from
the British Library.

Typeset by Rowland Phototypesetting Ltd, Bury St Edmunds, Suffolk
Printed and bound in the UK by CPI Mackays, Chatham, Kent ME5 8TD

Visit **www.panmacmillan.com** to read more about all our books
and to buy them. You will also find features, author interviews and
news of any author events, and you can sign up for e-newsletters
so that you're always first to hear about our new releases.

Contents

Editors' Foreword

The *2010 Yearbook* follows the long-established pattern, but with some important innovations. Wil Tirion, who produced our stars maps for the Northern and Southern Hemispheres, has this year produced all of the line diagrams showing the positions and movements of the planets to accompany the Monthly Notes. Martin Mobberley has provided the notes on eclipses, comets and minor planets, and Nick James has provided the data for the phases of the Moon, the longitudes of the Sun, Moon and planets, and details of lunar occultations. As always, John Isles and Bob Argyle have provided the information on variable stars and double stars respectively.

The articles section includes contributions both from our regular authors and from some very welcome newcomers. As usual, we have done our best to give you a wide range, both of subject and of technical level. For example, Martin Mobberley looks back a hundred years to the return of Halley's Comet in 1910 – the first return of the photographic age; David Rothery reviews the very latest exciting results obtained by the MESSENGER spacecraft during its close-range flybys of the planet Mercury; Fred Watson celebrates three dozen years of operations at the Anglo–Australian Telescope by looking back at the telescope's very significant achievements to date, and looking ahead to what the future may hold; and Iain Nicolson explores one of the great mysteries of modern cosmology – the extremely powerful gamma-ray bursts. For the more practically minded amateur astronomer, Paul Abel describes the techniques required to make useful visual observations of the planets; and Greg Parker shows how he has used the Hyperstar lens assembly on a Celestron Nexstar C11 Schmidt-Cassegrain telescope.

PATRICK MOORE
JOHN MASON
Selsey, August 2009

Preface

New readers will find that all the information in this *Yearbook* is given in diagrammatic or descriptive form; the positions of the planets may easily be found from the specially designed star charts, while the Monthly Notes describe the movements of the planets and give details of other astronomical phenomena visible in both the Northern and Southern Hemispheres. Two sets of star charts are provided. The **Northern Charts** (pp. 17 to 41) are designed for use at latitude 52°N, but may be used without alteration throughout the British Isles, and (except in the case of eclipses and occultations) in other countries of similar northerly latitude. The **Southern Charts** (pp. 43 to 67) are drawn for latitude 35°S, and are suitable for use in South Africa, Australia and New Zealand, and other locations in approximately the same southerly latitude. The reader who needs more detailed information will find *Norton's Star Atlas* an invaluable guide, while more precise positions of the planets and their satellites, together with predictions of occultations, meteor showers and periodic comets, may be found in the *Handbook of the British Astronomical Association*. Readers will also find details of forthcoming events given in the American monthly magazine *Sky & Telescope* and the British periodicals *The Sky at Night*, *Astronomy Now* and *Astronomy and Space*.

Important note

The times given on the star charts and in the Monthly Notes are generally given as local times, using the 24-hour clock, the day beginning at midnight. All the dates, and the times of a few events (e.g. eclipses) are given in Greenwich Mean Time (GMT), which is related to local time by the formula

Local Mean Time = GMT – west longitude

In practice, small differences in longitude are ignored, and the observer will use local clock time, which will be the appropriate Standard (or Zone) Time. As the formula indicates, places in west longitude will have a Standard Time slow on GMT, while places in east longitude will have a Standard Time fast on GMT. As examples we have:

Standard Time in

New Zealand	GMT + 12 hours
Victoria, NSW	GMT + 10 hours
Western Australia	GMT + 8 hours
South Africa	GMT + 2 hours
British Isles	GMT
Eastern ST	GMT − 5 hours
Central ST	GMT − 6 hours, etc.

If Summer Time is in use, the clocks will have been advanced by one hour, and this hour must be subtracted from the clock time to give Standard Time.

Part I

Monthly Charts and Astronomical Phenomena

Notes on the Star Charts

The stars, together with the Sun, Moon and planets, seem to be set on the surface of the celestial sphere, which appears to rotate about the Earth from east to west. Since it is impossible to represent a curved surface accurately on a plane, any kind of star map is bound to contain some form of distortion.

Most of the monthly star charts which appear in the various journals and some national newspapers are drawn in circular form. This is perfectly accurate, but it can make the charts awkward to use. For the star charts in this volume, we have preferred to give two hemispherical maps for each month of the year, one showing the northern aspect of the sky and the other showing the southern aspect. Two sets of monthly charts are provided, one for observers in the Northern Hemisphere and one for those in the Southern Hemisphere.

Unfortunately the constellations near the overhead point (the zenith) on these hemispherical charts can be rather distorted. This would be a serious drawback for precision charts, but what we have done is to give maps which are best suited to star recognition. We have also refrained from putting in too many stars, so that the main patterns stand out clearly. To help observers with any distortions near the zenith, and the lack of overlap between the charts of each pair, we have also included two circular maps, one showing all the constellations in the northern half of the sky, and one those in the southern half. Incidentally, there is a curious illusion that stars at an altitude of 60° or more are actually overhead, and beginners may often feel that they are leaning over backwards in trying to see them.

The charts show all stars down to the fourth magnitude, together with a number of fainter stars which are necessary to define the shapes of constellations. There is no standard system for representing the outlines of the constellations, and triangles and other simple figures have been used to give outlines which are easy to trace with the naked eye. The names of the constellations are given, together with the proper names of the brighter stars. The apparent magnitudes of the stars are indicated roughly by using different sizes of dot, the larger dots representing the brighter stars.

The two sets of star charts – one each for Northern and Southern Hemisphere observers – are similar in design. At each opening there is a single circular chart which shows all the constellations in that hemisphere of the sky. (These two charts are centred on the North and South Celestial Poles respectively.) Then there are twelve double-page spreads, showing the northern and southern aspects for each month of the year for observers in that hemisphere. In the **Northern Charts** (drawn for latitude 52°N) the left-hand chart of each spread shows the northern half of the sky (lettered 1N, 2N, 3N . . . 12N), and the corresponding right-hand chart shows the southern half of the sky (lettered 1S, 2S, 3S . . . 12S). The arrangement and lettering of the charts is exactly the same for the **Southern Charts** (drawn for latitude 35°S).

Because the sidereal day is shorter than the solar day, the stars appear to rise and set about four minutes earlier each day, and this amounts to two hours in a month. Hence the twelve pairs of charts in each set are sufficient to give the appearance of the sky throughout the day at intervals of two hours, or at the same time of night at monthly intervals throughout the year. For example, charts 1N and 1S here are drawn for 23 hours on 6 January. The view will also be the same on 6 October at 05 hours; 6 November at 03 hours; 6 December at 01 hours and 6 February at 21 hours. The actual range of dates and times when the stars on the charts are visible is indicated on each page. Each pair of charts is numbered in bold type, and the number to be used for any given month and time may be found from the following table:

Local Time	18ʰ	20ʰ	22ʰ	0ʰ	2ʰ	4ʰ	6ʰ
January	11	12	1	2	3	4	5
February	12	1	2	3	4	5	6
March	1	2	3	4	5	6	7
April	2	3	4	5	6	7	8
May	3	4	5	6	7	8	9
June	4	5	6	7	8	9	10
July	5	6	7	8	9	10	11
August	6	7	8	9	10	11	12
September	7	8	9	10	11	12	1

Local Time	18^h	20^h	22^h	0^h	2^h	4^h	6^h
October	8	9	10	11	12	1	2
November	9	10	11	12	1	2	3
December	10	11	12	1	2	3	4

On these charts, the ecliptic is drawn as a broken line on which longitude is marked every 10°. The positions of the planets are then easily found by reference to the table on p.74. It will be noticed that on the **Southern Charts** the ecliptic may reach an altitude in excess of 62.5° on the star charts showing the northern aspect (5N to 9N). The continuations of the broken line will be found on the corresponding charts for the southern aspect (5S, 6S, 8S and 9S).

Northern Star Charts

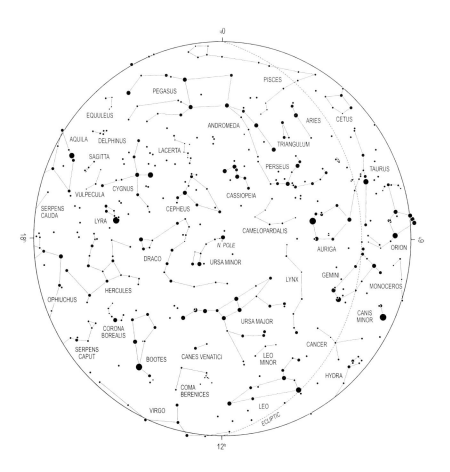

Northern Hemisphere

Note that the markers at 0ʰ, 6ʰ, 12ʰ and 18ʰ
indicate hours of Right Ascension.

1N

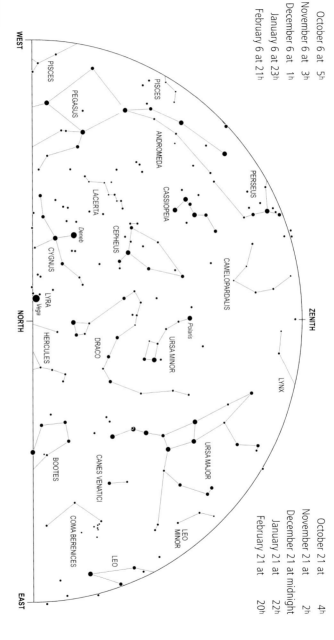

October 6 at 5h
November 6 at 3h
December 6 at 1h
January 6 at 23h
February 6 at 21h

WEST

PISCES
PEGASUS
PISCES
ANDROMEDA
PERSEUS
LACERTA
CASSIOPEIA
CEPHEUS
CAMELOPARDALIS
Deneb
CYGNUS
LYRA
Vega
NORTH
HERCULES
DRACO
Polaris
URSA MINOR
ZENITH
LYNX
BOOTES
CANES VENATICI
URSA MAJOR
LEO MINOR
COMA BERENICES
LEO
EAST

ZENITH

October 21 at 4h
November 21 at 2h
December 21 at midnight
January 21 at 22h
February 21 at 20h

1S

October 21 at 4ʰ
November 21 at 2ʰ
December 21 at midnight
January 21 at 22ʰ
February 21 at 20ʰ

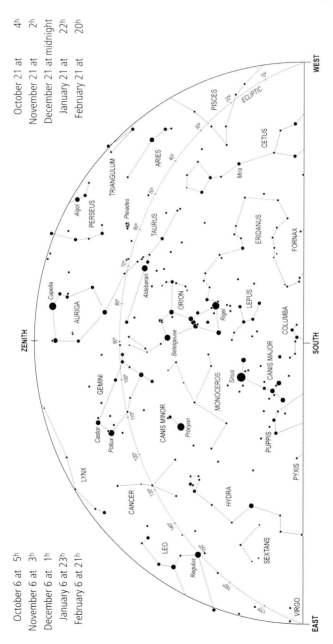

October 6 at 5ʰ
November 6 at 3ʰ
December 6 at 1ʰ
January 6 at 23ʰ
February 6 at 21ʰ

2N

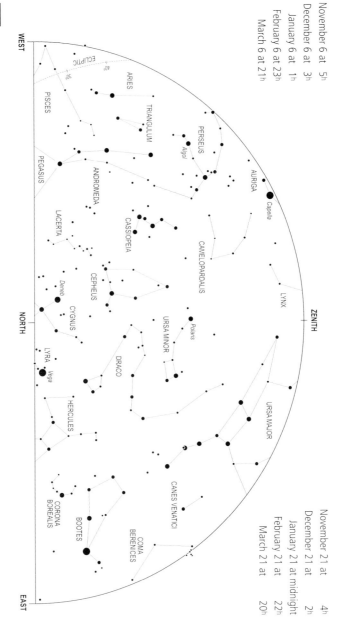

November 21 at   4ʰ
December 21 at   2ʰ
January 21 at midnight
February 21 at   22ʰ
March 21 at   20ʰ

WEST

NORTH

EAST

ZENITH

ECLIPTIC

ARIES
PISCES
TRIANGULUM
PERSEUS
Algol
AURIGA
Capella
PEGASUS
ANDROMEDA
CASSIOPEIA
CAMELOPARDALIS
LYNX
LACERTA
CEPHEUS
URSA MINOR
Polaris
Deneb
CYGNUS
DRACO
URSA MAJOR
LYRA
Vega
HERCULES
CANES VENATICI
CORONA BOREALIS
BOOTES
COMA BERENICES

2S

November 21 at   4ʰ
December 21 at   2ʰ
January 21 at midnight
February 21 at   22ʰ
March 21 at   20ʰ

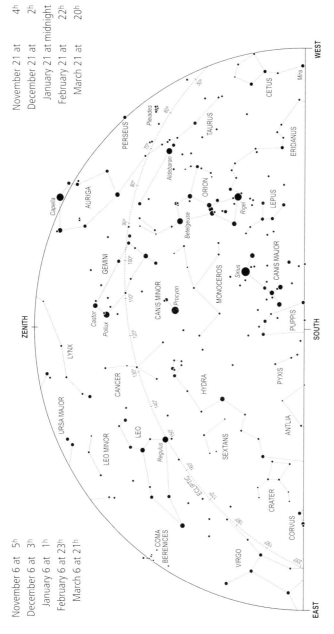

November 6 at   5ʰ
December 6 at   3ʰ
January 6 at   1ʰ
February 6 at 23ʰ
March 6 at 21ʰ

3N

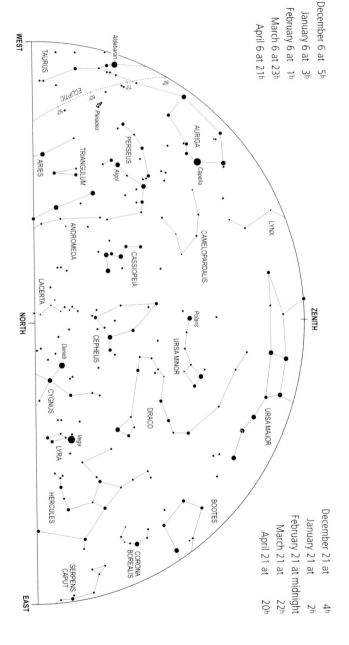

December 6 at 5h
January 6 at 3h
February 6 at 1h
March 6 at 23h
April 6 at 21h

WEST

TAURUS
Aldebaran
ECLIPTIC
Pleiades
ARIES
TRIANGULUM
PERSEUS
Algol
ANDROMEDA
AURIGA
Capella
LYNX
CAMELOPARDALIS
CASSIOPEIA
LACERTA
CEPHEUS
Polaris
URSA MINOR
ZENITH
NORTH
Deneb
CYGNUS
DRACO
URSA MAJOR
Vega
LYRA
HERCULES
BOOTES
CORONA
BOREALIS
SERPENS
CAPUT
EAST

December 21 at 4h
January 21 at 2h
February 21 at midnight
March 21 at 22h
April 21 at 20h

3S

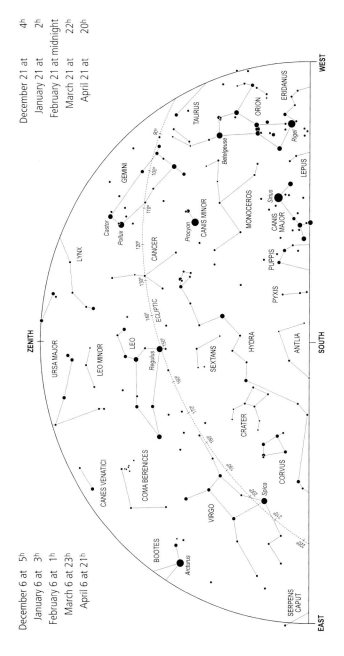

4N

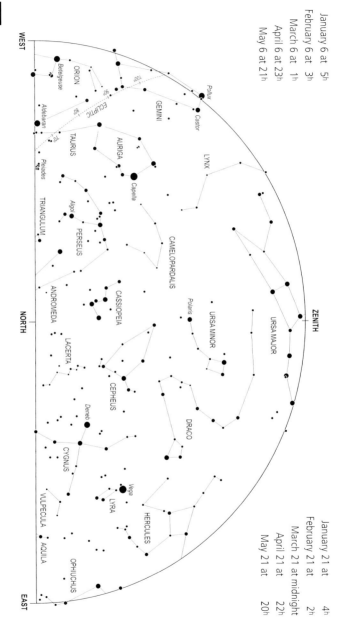

January 6 at 5h
February 6 at 3h
March 6 at 1h
April 6 at 23h
May 6 at 21h

January 21 at 4h
February 21 at 2h
March 21 at midnight
April 21 at 22h
May 21 at 20h

4S

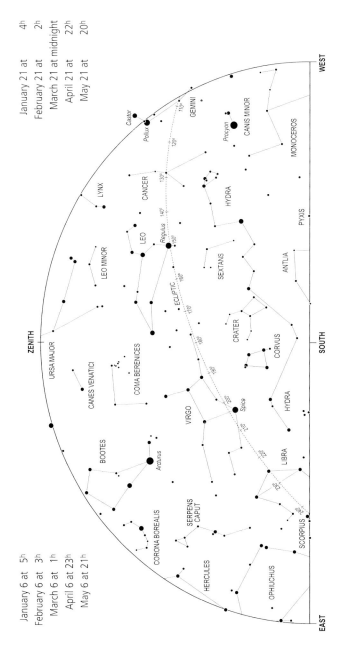

January 21 at   4h
February 21 at   2h
March 21 at midnight
April 21 at   22h
May 21 at   20h

January 6 at 5h
February 6 at 3h
March 6 at 1h
April 6 at 23h
May 6 at 21h

WEST
ZENITH
SOUTH
EAST

5N

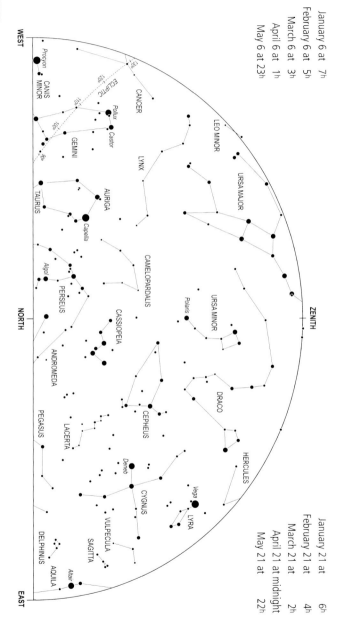

January 6 at 7h
February 6 at 5h
March 6 at 3h
April 6 at 1h
May 6 at 23h

January 21 at 6h
February 21 at 4h
March 21 at 2h
April 21 at midnight
May 21 at 22h

WEST

NORTH

EAST

ZENITH

Procyon
CANIS MINOR
CANCER
ECLIPTIC
Pollux
Castor
GEMINI
LEO MINOR
URSA MAJOR
LYNX
TAURUS
AURIGA
Capella
CAMELOPARDALIS
Algol
PERSEUS
Polaris
URSA MINOR
CASSIOPEIA
ANDROMEDA
DRACO
CEPHEUS
PEGASUS
LACERTA
HERCULES
Deneb
CYGNUS
Vega
LYRA
VULPECULA
SAGITTA
DELPHINUS
AQUILA
Altair

5S

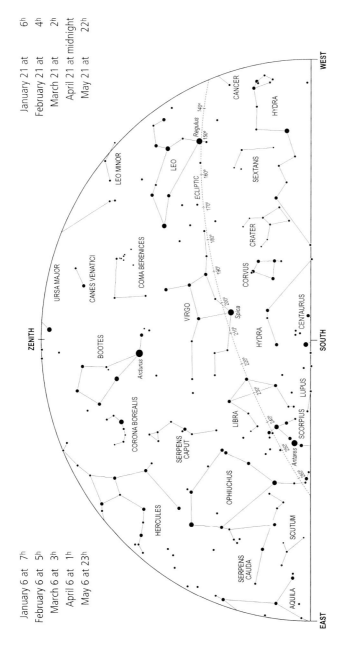

January 6 at 7ʰ
February 6 at 5ʰ
March 6 at 3ʰ
April 6 at 1ʰ
May 6 at 23ʰ

6N

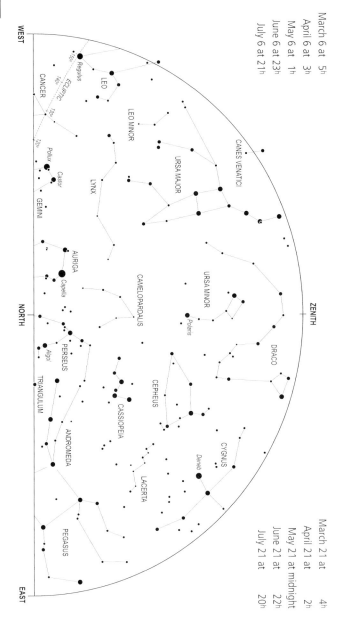

March 6 at 5h
April 6 at 3h
May 6 at 1h
June 6 at 23h
July 6 at 21h

March 21 at 4h
April 21 at 2h
May 21 at midnight
June 21 at 22h
July 21 at 20h

6S

March 21 at 4ʰ
April 21 at 2ʰ
May 21 at midnight
June 21 at 22ʰ
July 21 at 20ʰ

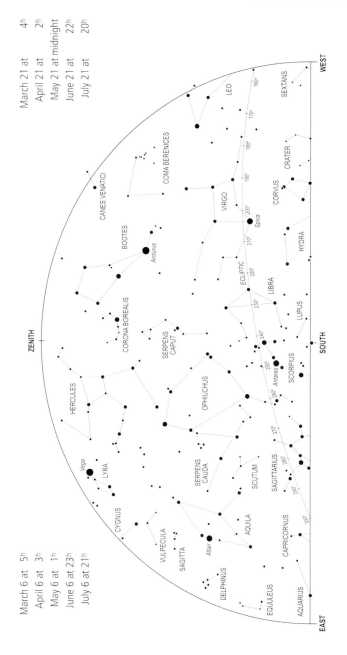

March 6 at 5ʰ
April 6 at 3ʰ
May 6 at 1ʰ
June 6 at 23ʰ
July 6 at 21ʰ

7N

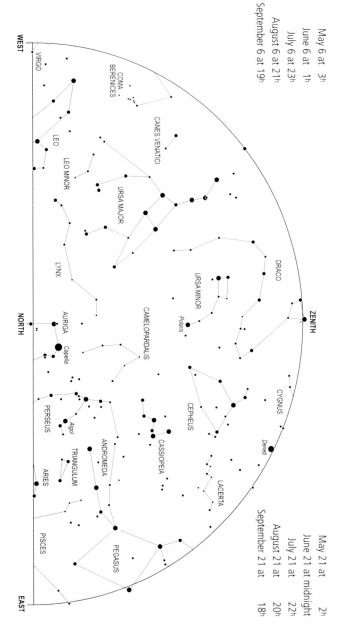

WEST

May 6 at 3h
June 6 at 1h
July 6 at 23h
August 6 at 21h
September 6 at 19h

VIRGO
COMA BERENICES
CANES VENATICI
LEO
LEO MINOR
URSA MAJOR
LYNX
CANCER
DRACO
URSA MINOR
Polaris
CAMELOPARDALIS
AURIGA
Capella
CEPHEUS
CYGNUS
ZENITH
Deneb

NORTH

PERSEUS
Algol
ANDROMEDA
CASSIOPEIA
TRIANGULUM
LACERTA
ARIES
PISCES
PEGASUS

EAST

May 21 at 2h
June 21 at midnight
July 21 at 22h
August 21 at 20h
September 21 at 18h

7S

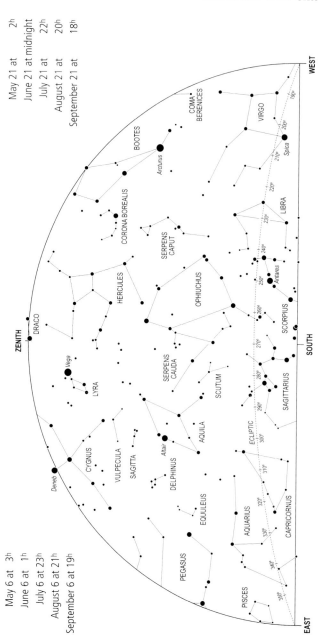

May 21 at   2ʰ
June 21 at midnight
July 21 at   22ʰ
August 21 at   20ʰ
September 21 at 18ʰ

May 6 at   3ʰ
June 6 at   1ʰ
July 6 at   23ʰ
August 6 at 21ʰ
September 6 at 19ʰ

WEST

ZENITH

SOUTH

EAST

COMA BERENICES
BOOTES
Arcturus
VIRGO
Spica
CORONA BOREALIS
SERPENS CAPUT
LIBRA
HERCULES
OPHIUCHUS
Antares
DRACO
SCORPIUS
Vega
LYRA
SERPENS CAUDA
SCUTUM
SAGITTARIUS
CYGNUS
VULPECULA
SAGITTA
Altair
AQUILA
Deneb
DELPHINUS
ECLIPTIC
EQUULEUS
AQUARIUS
CAPRICORNUS
PEGASUS
PISCES

180°
200°
210°
220°
230°
240°
250°
260°
270°
280°
290°
300°
310°
320°
330°
340°
350°

8N

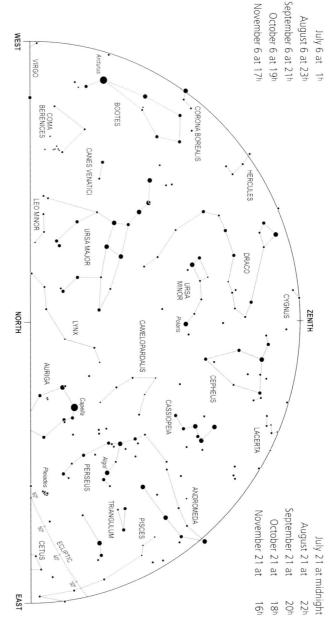

July 6 at 1h
August 6 at 23h
September 6 at 21h
October 6 at 19h
November 6 at 17h

July 21 at midnight
August 21 at 22h
September 21 at 20h
October 21 at 18h
November 21 at 16h

8S

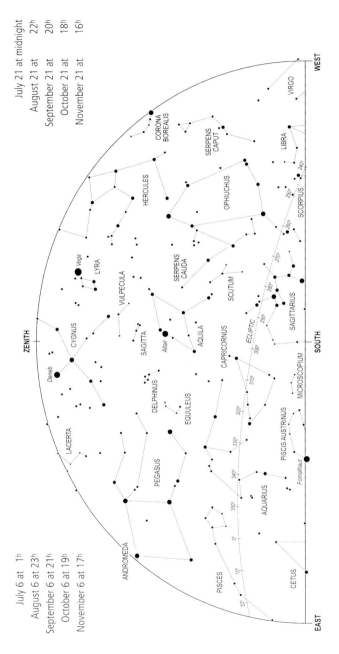

July 21 at midnight 22ʰ
August 21 at 22ʰ
September 21 at 20ʰ
October 21 at 18ʰ
November 21 at 16ʰ

July 6 at 1ʰ
August 6 at 23ʰ
September 6 at 21ʰ
October 6 at 19ʰ
November 6 at 17ʰ

WEST

VIRGO

CORONA BOREALIS

SERPENS CAPUT

LIBRA

HERCULES

OPHIUCHUS

SCORPIUS

24ᵖ
25ᵖ
26ᵖ

ZENITH

Vega
LYRA

VULPECULA

SERPENS CAUDA

SCUTUM

270ᵖ

CYGNUS

SAGITTA

Altair

AQUILA

280ᵖ

SAGITTARIUS

290ᵖ

Deneb

CAPRICORNUS

ECLIPTIC

300ᵖ

SOUTH

LACERTA

DELPHINUS

EQUULEUS

310ᵖ

MICROSCOPIUM

320ᵖ

PEGASUS

330ᵖ

PISCIS AUSTRINUS

ANDROMEDA

340ᵖ

350ᵖ

Fomalhaut

0ᵖ

AQUARIUS

PISCES

10ᵖ

CETUS

20ᵖ

EAST

9N

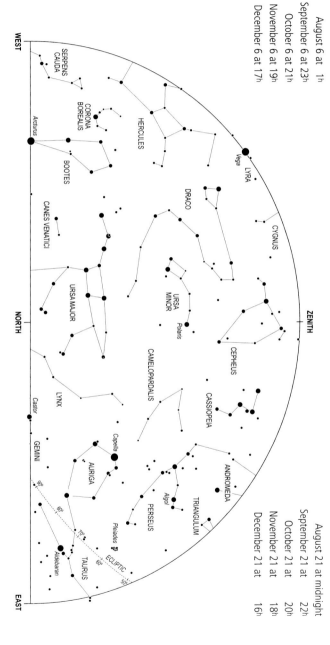

August 6 at 1h
September 6 at 23h
October 6 at 21h
November 6 at 19h
December 6 at 17h

August 21 at midnight
September 21 at 22h
October 21 at 20h
November 21 at 18h
December 21 at 16h

9S

August 21 at midnight
September 21 at   22ʰ
October 21 at   20ʰ
November 21 at   18ʰ
December 21 at   16ʰ

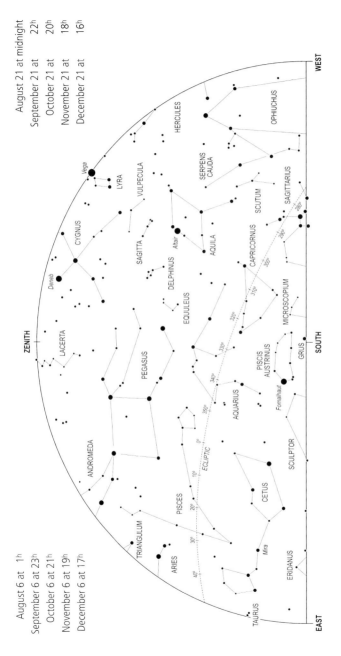

August 6 at   1ʰ
September 6 at 23ʰ
October 6 at 21ʰ
November 6 at 19ʰ
December 6 at 17ʰ

10N

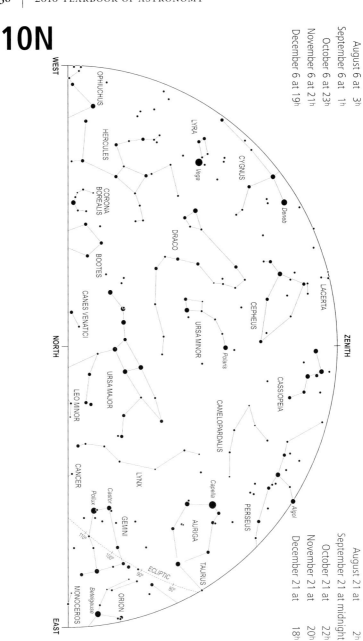

August 6 at 3h
September 6 at 1h
October 6 at 23h
November 6 at 21h
December 6 at 19h

WEST

OPHIUCHUS
HERCULES
CORONA BOREALIS
BOOTES
CANES VENATICI
LEO MINOR
URSA MAJOR
CANCER
LYNX
MONOCEROS
ORION
Betelgeuse
GEMINI
Castor
Pollux
LYRA
Vega
CYGNUS
Deneb
DRACO
URSA MINOR
Polaris
CEPHEUS
LACERTA
CASSIOPEIA
CAMELOPARDALIS
PERSEUS
Algol
Capella
AURIGA
TAURUS
ECLIPTIC
80°
90°
100°
110°

NORTH

ZENITH

EAST

August 21 at 2h
September 21 at midnight
October 21 at 22h
November 21 at 20h
December 21 at 18h

10S

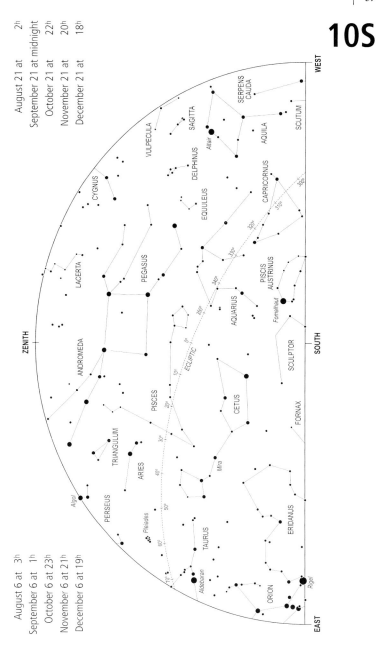

August 21 at 2ʰ
September 21 at midnight
October 21 at 22ʰ
November 21 at 20ʰ
December 21 at 18ʰ

August 6 at 3ʰ
September 6 at 1ʰ
October 6 at 23ʰ
November 6 at 21ʰ
December 6 at 19ʰ

WEST

SERPENS CAUDA
SAGITTA
VULPECULA
SCUTUM
Altair
AQUILA
DELPHINUS
CYGNUS
CAPRICORNUS
EQUULEUS
30°
31°
LACERTA
32°
PEGASUS
33°
PISCIS AUSTRINUS
ZENITH
34°
AQUARIUS
Fomalhaut
35°
ANDROMEDA
0°
SOUTH
ECLIPTIC
10°
SCULPTOR
PISCES
20°
CETUS
FORNAX
TRIANGULUM
30°
ARIES
40°
Mira
Algol
PERSEUS
50°
ERIDANUS
Pleiades
60°
TAURUS
70°
Aldebaran
ORION
Rigel
EAST

11N

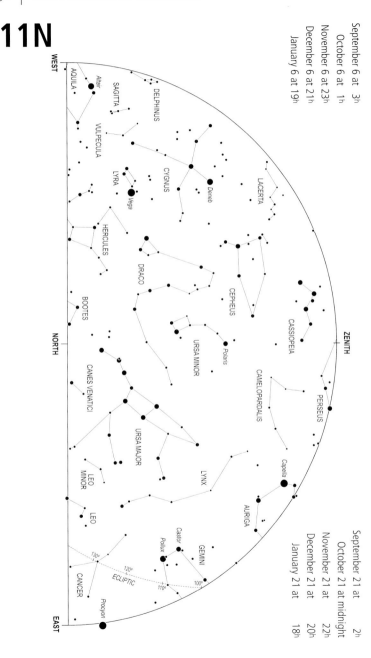

September 6 at 3h
October 6 at 1h
November 6 at 23h
December 6 at 21h
January 6 at 19h

September 21 at 2h
October 21 at midnight
November 21 at 22h
December 21 at 20h
January 21 at 18h

11S

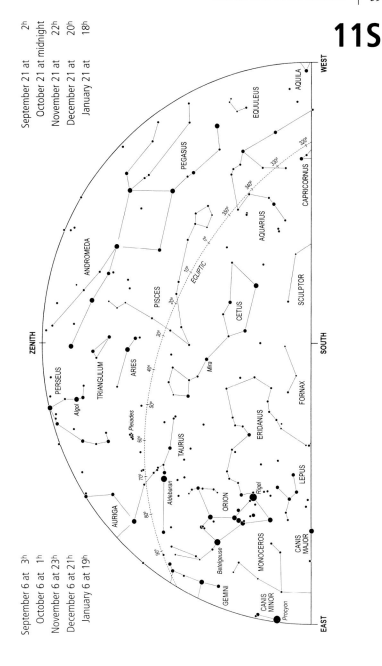

September 21 at 2h
October 21 at midnight
November 21 at 22h
December 21 at 20h
January 21 at 18h

September 6 at 3h
October 6 at 1h
November 6 at 23h
December 6 at 21h
January 6 at 19h

WEST

ZENITH

SOUTH

EAST

AQUILA
EQUULEUS
PEGASUS
ANDROMEDA
PISCES
AQUARIUS
CAPRICORNUS
SCULPTOR
CETUS
ECLIPTIC
Mira
TRIANGULUM
ARIES
PERSEUS
Algol
Pleiades
TAURUS
FORNAX
ERIDANUS
Aldebaran
ORION
Rigel
LEPUS
AURIGA
Betelgeuse
MONOCEROS
CANIS MAJOR
GEMINI
CANIS MINOR
Procyon

320°
330°
340°
350°
0°
10°
20°
30°
40°
50°
60°
70°
80°
90°

12N

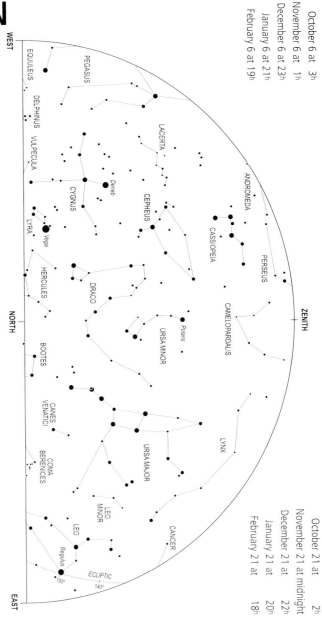

October 6 at 3h
November 6 at 1h
December 6 at 23h
January 6 at 21h
February 6 at 19h

WEST

EQUULEUS
PEGASUS
DELPHINUS
VULPECULA
LACERTA
CYGNUS
Deneb
CEPHEUS
ANDROMEDA
LYRA
Vega
CASSIOPEIA
PERSEUS
HERCULES
DRACO
CAMELOPARDALIS
BOOTES
Polaris
URSA MINOR
ZENITH
CANES VENATICI
LYNX
COMA BERENICES
URSA MAJOR
LEO MINOR
CANCER
LEO
Regulus
ECLIPTIC
150°   140°

NORTH

EAST

October 21 at 2h
November 21 at midnight
December 21 at 22h
January 21 at 20h
February 21 at 18h

12S

October 21 at   2h
November 21 at midnight
December 21 at   22h
January 21 at   20h
February 21 at   18h

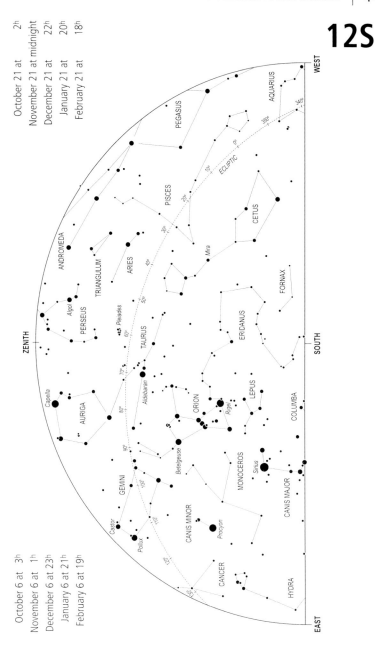

October 6 at   3h
November 6 at   1h
December 6 at 23h
January 6 at   21h
February 6 at 19h

Southern Star Charts

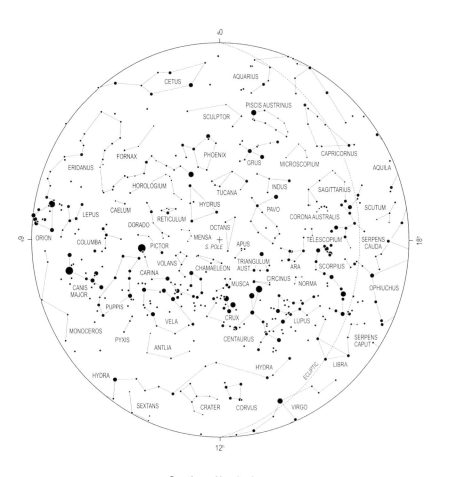

Southern Hemisphere

Note that the markers at 0ʰ, 6ʰ, 12ʰ and 18ʰ
indicate hours of Right Ascension.

1N

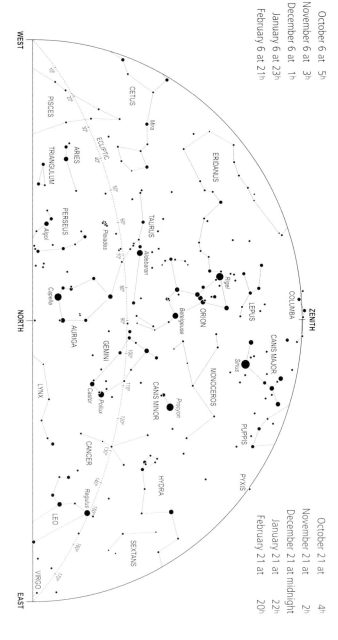

1S

October 21 at 4ʰ
November 21 at 2ʰ
December 21 at midnight
January 21 at 22ʰ
February 21 at 20ʰ

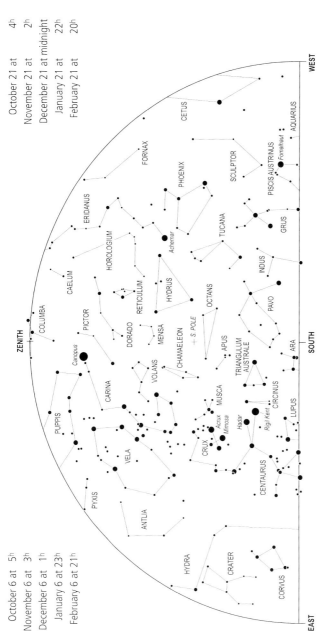

October 6 at 5ʰ
November 6 at 3ʰ
December 6 at 1ʰ
January 6 at 23ʰ
February 6 at 21ʰ

2N

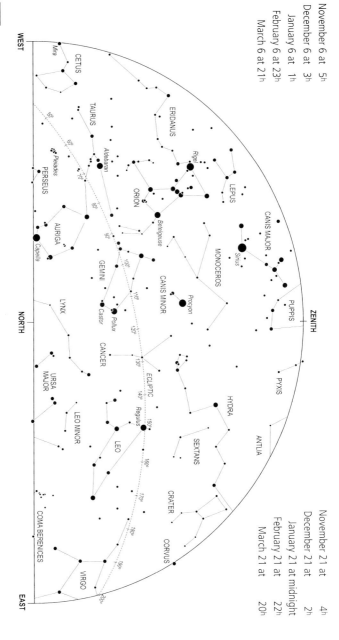

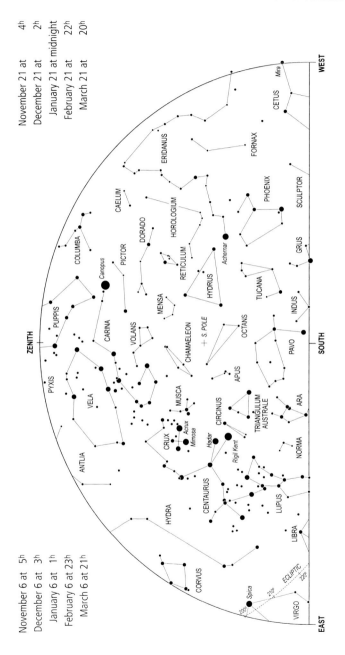

November 21 at   4ʰ
December 21 at   2ʰ
January 21 at midnight
February 21 at   22ʰ
March 21 at   20ʰ

WEST

Mira

CETUS

ERIDANUS

FORNAX

CAELUM

PHOENIX

SCULPTOR

DORADO

HOROLOGIUM

COLUMBA

PICTOR

RETICULUM

Achernar

GRUS

Canopus

MENSA

HYDRUS

TUCANA

PUPPIS

CARINA

VOLANS

INDUS

ZENITH

CHAMAELEON

+ S. POLE

OCTANS

PYXIS

VELA

MUSCA

APUS

PAVO

SOUTH

CRUX

CIRCINUS

Acrux

Mimosa

Hadar

TRIANGULUM
AUSTRALE

ARA

ANTLIA

Rigil Kent

NORMA

CENTAURUS

LUPUS

HYDRA

LIBRA

CORVUS

Spica

210°

ECLIPTIC

220°

200°

VIRGO

EAST

November 6 at   5ʰ
December 6 at   3ʰ
January 6 at   1ʰ
February 6 at   23ʰ
March 6 at   21ʰ

3N

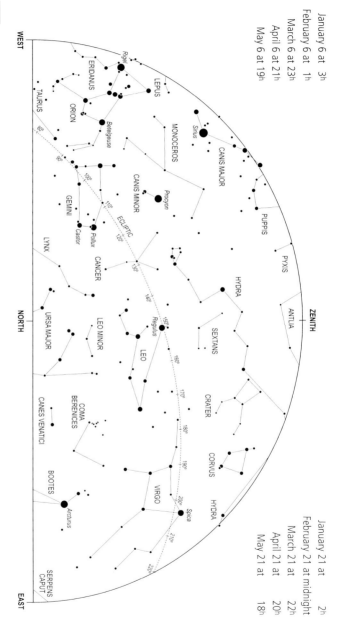

January 6 at 3h
February 6 at 1h
March 6 at 23h
April 6 at 21h
May 6 at 19h

January 21 at 2h
February 21 at midnight
March 21 at 22h
April 21 at 20h
May 21 at 18h

3S

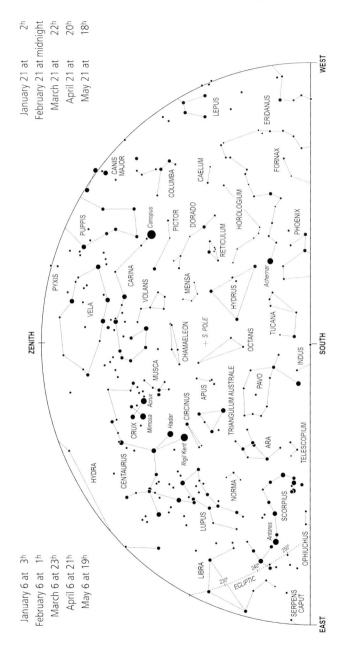

4N

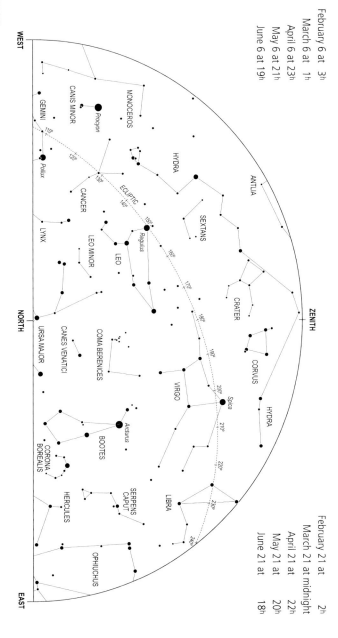

4S

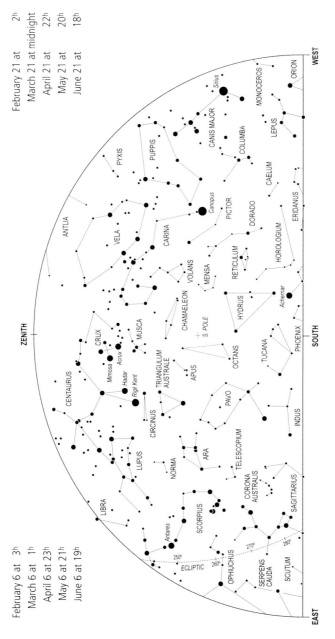

WEST

ZENITH

SOUTH

EAST

ORION

MONOCEROS

LEPUS

CANIS MAJOR

Sirius

COLUMBA

CAELUM

ERIDANUS

PUPPIS

PYXIS

CARINA

PICTOR

DORADO

HOROLOGIUM

Canopus

RETICULUM

ANTLIA

VELA

VOLANS

MENSA

HYDRUS

Achernar

CHAMAELEON

S. POLE

PHOENIX

CRUX

MUSCA

Acrux

Mimosa

OCTANS

TUCANA

CENTAURUS

Hadar

Rigil Kent

TRIANGULUM
AUSTRALE

APUS

PAVO

INDUS

CIRCINUS

LUPUS

NORMA

ARA

TELESCOPIUM

LIBRA

Antares

SCORPIUS

CORONA
AUSTRALIS

SAGITTARIUS

250°

270°

260°

OPHIUCHUS

260°

SERPENS
CAUDA

SCUTUM

ECLIPTIC

5N

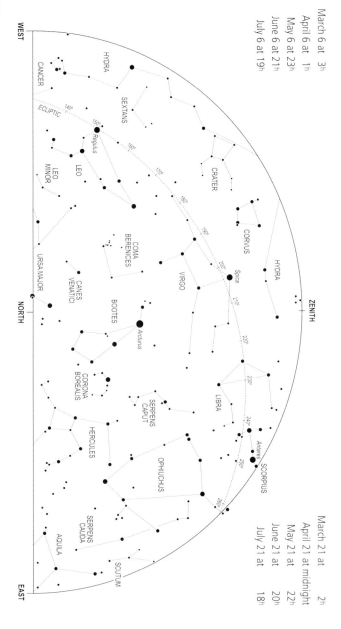

March 6 at 3ʰ
April 6 at 1ʰ
May 6 at 23ʰ
June 6 at 21ʰ
July 6 at 19ʰ

March 21 at 2ʰ
April 21 at midnight
May 21 at 22ʰ
June 21 at 20ʰ
July 21 at 18ʰ

5S

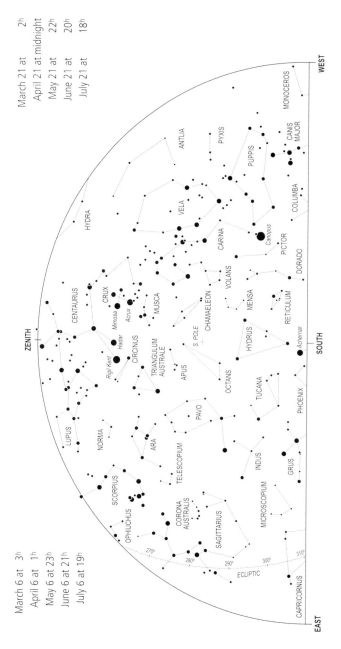

WEST

March 21 at 2ʰ
April 21 at midnight
May 21 at 22ʰ
June 21 at 20ʰ
July 21 at 18ʰ

MONOCEROS

CANIS MAJOR

PUPPIS

COLUMBA

ANTLIA

PYXIS

VELA

CARINA

Canopus

PICTOR

DORADO

HYDRA

VOLANS

CENTAURUS

CRUX

Mimosa

Acrux

MUSCA

CHAMAELEON

MENSA

RETICULUM

Achernar

ZENITH

Hadar

CIRCINUS

TRIANGULUM AUSTRALE

S. POLE

HYDRUS

Rigil Kent

APUS

OCTANS

TUCANA

SOUTH

PHOENIX

LUPUS

NORMA

PAVO

ARA

TELESCOPIUM

INDUS

GRUS

SCORPIUS

OPHIUCHUS

CORONA AUSTRALIS

SAGITTARIUS

MICROSCOPIUM

270° 280° 290° 300° 310°

ECLIPTIC

CAPRICORNUS

EAST

March 6 at 3ʰ
April 6 at 1ʰ
May 6 at 23ʰ
June 6 at 21ʰ
July 6 at 19ʰ

6N

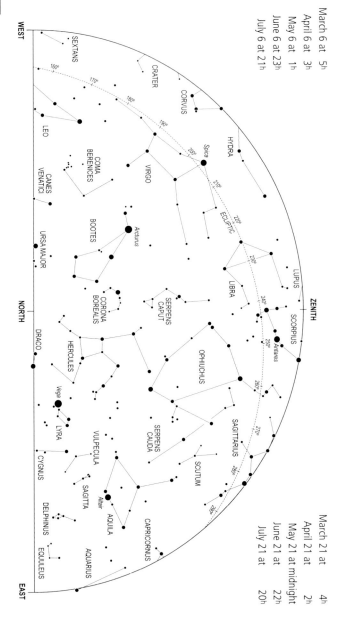

March 6 at 5h
April 6 at 3h
May 6 at 1h
June 6 at 23h
July 6 at 21h

March 21 at 4h
April 21 at 2h
May 21 at midnight
June 21 at 22h
July 21 at 20h

WEST

SEXTANS
CRATER
CORVUS
HYDRA
LEO
COMA BERENICES
CANES VENATICI
VIRGO
Spica
LUPUS
ZENITH
URSA MAJOR
BOOTES
Arcturus
LIBRA
SCORPIUS
Antares
NORTH
CORONA BOREALIS
SERPENS CAPUT
OPHIUCHUS
DRACO
HERCULES
ECLIPTIC
Vega
LYRA
SERPENS CAUDA
SAGITTARIUS
CYGNUS
VULPECULA
SCUTUM
DELPHINUS
SAGITTA
Altair
AQUILA
CAPRICORNUS
EQUULEUS
AQUARIUS

EAST

160° 170° 180° 190° 200° 210° 220° 230° 240° 250° 260° 270° 280° 290°

6S

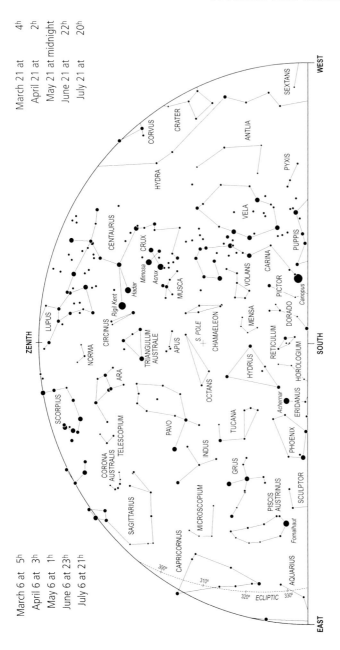

March 21 at   4ʰ
April 21 at   2ʰ
May 21 at midnight
June 21 at   22ʰ
July 21 at   20ʰ

March 6 at   5ʰ
April 6 at   3ʰ
May 6 at   1ʰ
June 6 at   23ʰ
July 6 at   21ʰ

WEST

SEXTANS
CORVUS
CRATER
ANTLIA
HYDRA
PYXIS
VELA
PUPPIS
CENTAURUS
CRUX
CARINA
Mimosa
Acrux
MUSCA
VOLANS
PICTOR
Hadar
Rigil Kent
CIRCINUS
S. POLE
CHAMAELEON
MENSA
DORADO
Canopus
RETICULUM
LUPUS
TRIANGULUM
AUSTRALE
APUS
HOROLOGIUM
ZENITH
NORMA
ARA
OCTANS
HYDRUS
Achernar
ERIDANUS
SOUTH
SCORPIUS
PAVO
TUCANA
PHOENIX
TELESCOPIUM
INDUS
CORONA
AUSTRALIS
GRUS
PISCIS
AUSTRINUS
SCULPTOR
SAGITTARIUS
MICROSCOPIUM
Fomalhaut
CAPRICORNUS
300°
310°
320° ECLIPTIC 330°
AQUARIUS

EAST

7N

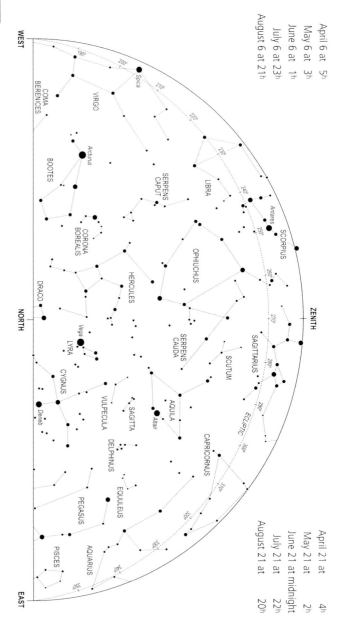

April 21 at 4h
May 21 at 2h
June 21 at midnight
July 21 at 22h
August 21 at 20h

7S

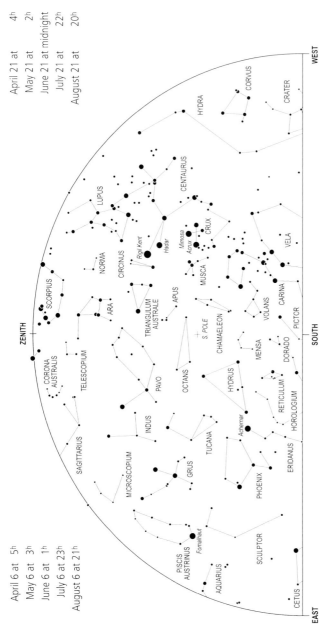

April 21 at   4ʰ
May 21 at   2ʰ
June 21 at midnight
July 21 at   22ʰ
August 21 at   20ʰ

April 6 at   5ʰ
May 6 at   3ʰ
June 6 at   1ʰ
July 6 at   23ʰ
August 6 at   21ʰ

WEST

EAST

ZENITH

SOUTH

CORVUS
CRATER
HYDRA
CENTAURUS
LUPUS
CRUX
Mimosa
Acrux
Hadar
Rigil Kent
MUSCA
VELA
NORMA
CIRCINUS
ARA
APUS
TRIANGULUM
AUSTRALE
MUSCA
CARINA
VOLANS
PICTOR
SCORPIUS
CHAMAELEON
S. POLE
MENSA
DORADO
CORONA
AUSTRALIS
TELESCOPIUM
PAVO
OCTANS
HYDRUS
RETICULUM
HOROLOGIUM
INDUS
Achernar
ERIDANUS
SAGITTARIUS
TUCANA
MICROSCOPIUM
GRUS
PHOENIX
SCULPTOR
PISCIS
AUSTRINUS
Fomalhaut
AQUARIUS
CETUS

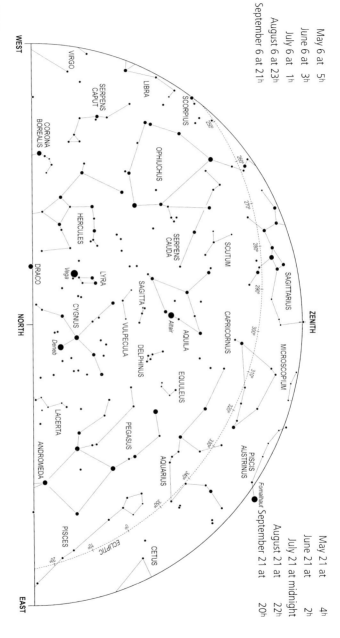

8N

May 6 at 5h
June 6 at 3h
July 6 at 1h
August 6 at 23h
September 6 at 21h

May 21 at 4h
June 21 at 2h
July 21 at midnight
August 21 at 22h
September 21 at 20h

WEST

NORTH

EAST

ZENITH

VIRGO
SERPENS CAPUT
CORONA BOREALIS
LIBRA
SCORPIUS
OPHIUCHUS
HERCULES
SERPENS CAUDA
SCUTUM
SAGITTARIUS
DRACO
Vega
LYRA
SAGITTA
CAPRICORNUS
MICROSCOPIUM
CYGNUS
Altair
AQUILA
VULPECULA
Deneb
DELPHINUS
EQUULEUS
AQUARIUS
PISCIS AUSTRINUS
LACERTA
PEGASUS
Fomalhaut
ANDROMEDA
AQUARIUS
PISCES
ECLIPTIC
CETUS

8S

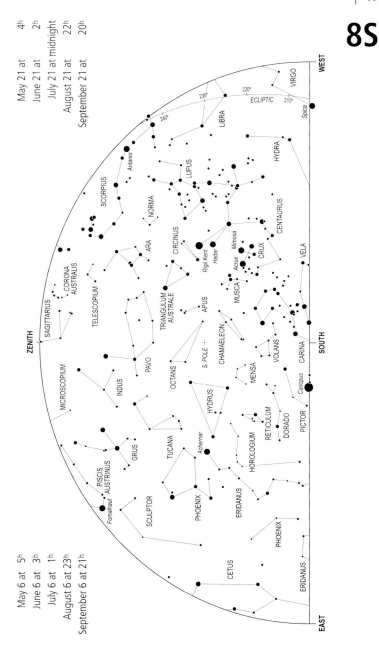

WEST

ZENITH

SOUTH

EAST

VIRGO
Spica
ECLIPTIC 210°
220°
230°
240°
LIBRA
HYDRA
LUPUS
Antares
SCORPIUS
NORMA
CENTAURUS
ARA
CIRCINUS
Rigil Kent
Hadar
Mimosa
Acrux
CRUX
VELA
CORONA AUSTRALIS
SAGITTARIUS
TELESCOPIUM
TRIANGULUM AUSTRALE
APUS
MUSCA
CHAMAELEON
MICROSCOPIUM
PAVO
S. POLE
OCTANS
VOLANS
CARINA
Canopus
INDUS
HYDRUS
MENSA
RETICULUM
DORADO
PICTOR
GRUS
TUCANA
Achernar
HOROLOGIUM
PISCIS AUSTRINUS
Fomalhaut
SCULPTOR
PHOENIX
ERIDANUS
PHOENIX
CETUS
ERIDANUS

9N

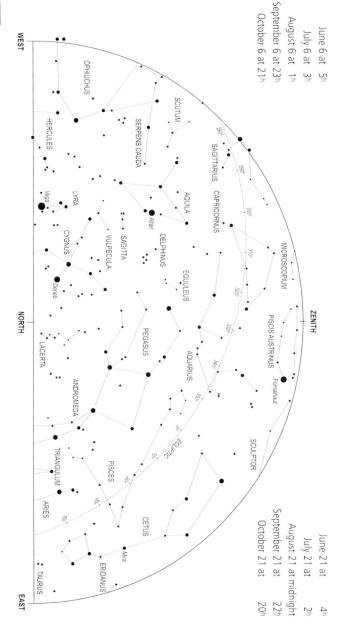

June 6 at 5h
July 6 at 3h
August 6 at 1h
September 6 at 23h
October 6 at 21h

June 21 at 4h
July 21 at 2h
August 21 at midnight
September 21 at 22h
October 21 at 20h

WEST

NORTH

EAST

ZENITH

OPHIUCHUS
SCUTUM
SERPENS CAUDA
SAGITTARIUS
CAPRICORNUS
MICROSCOPIUM
HERCULES
AQUILA
Altair
LYRA
Vega
CYGNUS
DELPHINUS
SAGITTA
VULPECULA
Deneb
EQUULEUS
PISCIS AUSTRINUS
Fomalhaut
AQUARIUS
PEGASUS
SCULPTOR
LACERTA
ANDROMEDA
TRIANGULUM
PISCES
ARIES
CETUS
Mira
ERIDANUS
TAURUS
ECLIPTIC

280°
290°
300°
310°
320°
330°
340°
350°
0°
10°
20°
30°
40°

9S

June 21 at 4ʰ
July 21 at 2ʰ
August 21 at midnight
September 21 at 22ʰ
October 21 at 20ʰ

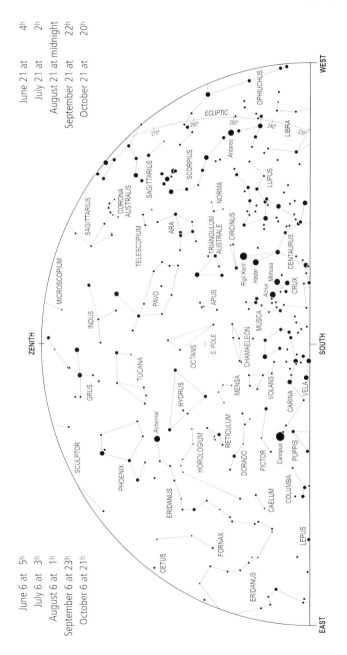

ECLIPTIC

WEST

ZENITH

SOUTH

EAST

June 6 at 5ʰ
July 6 at 3ʰ
August 6 at 1ʰ
September 6 at 23ʰ
October 6 at 21ʰ

OPHIUCHUS
LIBRA
Antares
SCORPIUS
LUPUS
SAGITTARIUS
CORONA AUSTRALIS
NORMA
CENTAURUS
TRIANGULUM AUSTRALE
CIRCINUS
ARA
TELESCOPIUM
Rigil Kent
Hadar
Acrux
Mimosa
CRUX
APUS
PAVO
MUSCA
MICROSCOPIUM
INDUS
CHAMAELEON
MENSA
VOLANS
VELA
OCTANS
S. POLE
TUCANA
CARINA
HYDRUS
GRUS
Achernar
RETICULUM
DORADO
PICTOR
Canopus
PUPPIS
SCULPTOR
PHOENIX
HOROLOGIUM
CAELUM
COLUMBA
ERIDANUS
FORNAX
LEPUS
CETUS
ERIDANUS

270° 260° 250° 240° 230°

10N

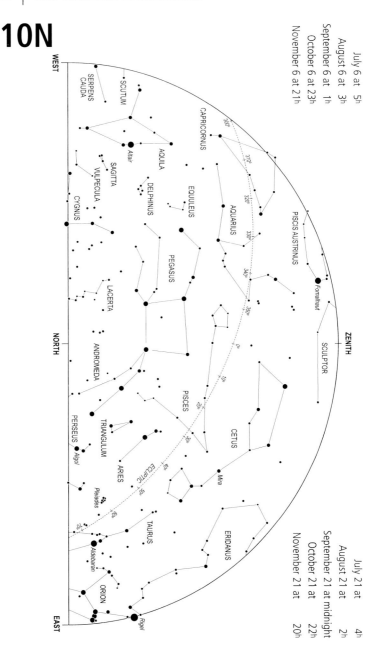

July 21 at 4h
August 21 at 2h
September 21 at midnight
October 21 at 22h
November 21 at 20h

10S

July 21 at 4ʰ
August 21 at 2ʰ
September 21 at midnight
October 21 at 22ʰ
November 21 at 20ʰ

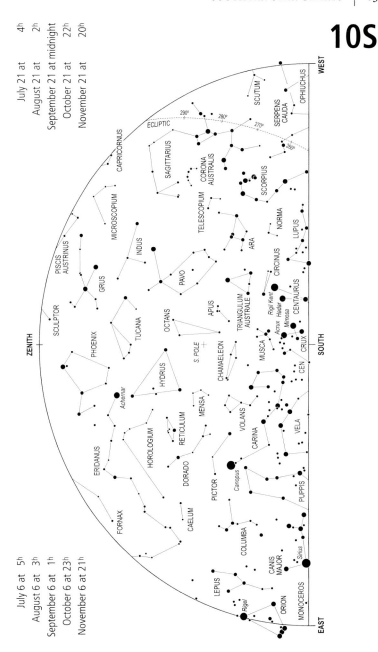

July 6 at 5ʰ
August 6 at 3ʰ
September 6 at 1ʰ
October 6 at 23ʰ
November 6 at 21ʰ

11N

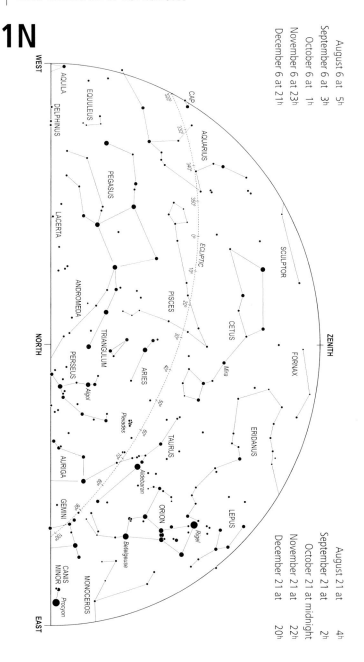

August 6 at 5h
September 6 at 3h
October 6 at 1h
November 6 at 23h
December 6 at 21h

WEST

August 21 at 4h
September 21 at 2h
October 21 at midnight
November 21 at 22h
December 21 at 20h

EAST

NORTH

ZENITH

AQUILA
DELPHINUS
EQUULEUS
PEGASUS
LACERTA
ANDROMEDA
TRIANGULUM
PERSEUS
Algol
AURIGA
GEMINI
CANIS MINOR
Procyon
MONOCEROS
Betelgeuse
ORION
Rigel
LEPUS
TAURUS
Aldebaran
Pleiades
ARIES
Mira
CETUS
PISCES
ECLIPTIC
AQUARIUS
CAP
SCULPTOR
FORNAX
ERIDANUS

11S

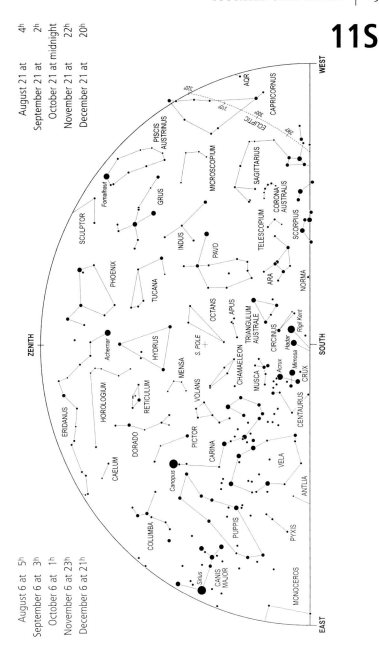

August 21 at 4ʰ
September 21 at 2ʰ
October 21 at midnight
November 21 at 22ʰ
December 21 at 20ʰ

August 6 at 5ʰ
September 6 at 3ʰ
October 6 at 1ʰ
November 6 at 23ʰ
December 6 at 21ʰ

WEST

ZENITH

SOUTH

EAST

AQR
CAPRICORNUS
ECLIPTIC
320°
310°
300°
290°
PISCIS AUSTRINUS
MICROSCOPIUM
SAGITTARIUS
CORONA AUSTRALIS
GRUS
Fomalhaut
SCULPTOR
INDUS
PAVO
TELESCOPIUM
SCORPIUS
NORMA
ARA
PHOENIX
TUCANA
OCTANS
APUS
TRIANGULUM AUSTRALE
CIRCINUS
Rigil Kent
Hadar
Mimosa
Acrux
CRUX
Achernar
HYDRUS
S. POLE
CHAMAELEON
MUSCA
CENTAURUS
ERIDANUS
HOROLOGIUM
RETICULUM
MENSA
VOLANS
CARINA
VELA
CAELUM
DORADO
PICTOR
Canopus
ANTLIA
COLUMBA
PUPPIS
PYXIS
Sirius
CANIS MAJOR
MONOCEROS

12N

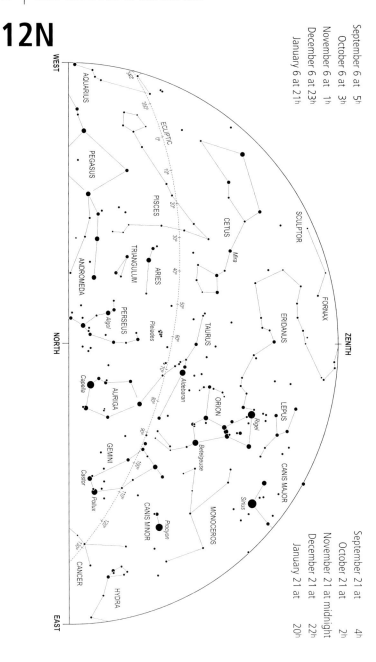

September 6 at 5h
October 6 at 3h
November 6 at 1h
December 6 at 23h
January 6 at 21h

September 21 at 4h
October 21 at 2h
November 21 at midnight
December 21 at 22h
January 21 at 20h

12S

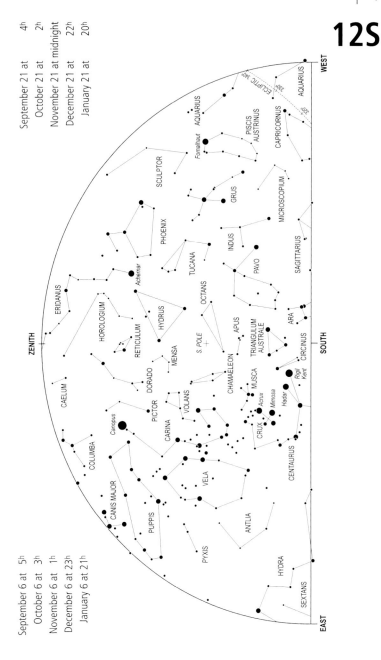

September 21 at 4ʰ
October 21 at 2ʰ
November 21 at midnight
December 21 at 22ʰ
January 21 at 20ʰ

September 6 at 5ʰ
October 6 at 3ʰ
November 6 at 1ʰ
December 6 at 23ʰ
January 6 at 21ʰ

WEST

ZENITH

SOUTH

EAST

AQUARIUS
ECLIPTIC 340°
330°
320°
AQUARIUS
PISCIS AUSTRINUS
CAPRICORNUS
Fomalhaut
MICROSCOPIUM
SCULPTOR
GRUS
SAGITTARIUS
PHOENIX
INDUS
TUCANA
PAVO
Achernar
OCTANS
ERIDANUS
ARA
HOROLOGIUM
HYDRUS
APUS
TRIANGULUM AUSTRALE
RETICULUM
S. POLE
CIRCINUS
MENSA
CHAMAELEON
Rigil Kent
CAELUM
DORADO
MUSCA
Hadar
PICTOR
VOLANS
Acrux
Mimosa
CRUX
Canopus
CARINA
CENTAURUS
COLUMBA
CANIS MAJOR
VELA
PUPPIS
ANTLIA
PYXIS
HYDRA
SEXTANS

The Planets and the Ecliptic

The paths of the planets about the Sun all lie close to the plane of the ecliptic, which is marked for us in the sky by the apparent path of the Sun among the stars, and is shown on the star charts by a broken line. The Moon and naked-eye planets will always be found close to this line, never departing from it by more than about 7°. Thus the planets are most favourably placed for observation when the ecliptic is well displayed, and this means that it should be as high in the sky as possible. This avoids the difficulty of finding a clear horizon, and also overcomes the problem of atmospheric absorption, which greatly reduces the light of the stars. Thus a star at an altitude of 10° suffers a loss of 60 per cent of its light, which corresponds to a whole magnitude; at an altitude of only 4°, the loss may amount to two magnitudes.

The position of the ecliptic in the sky is therefore of great importance, and since it is tilted at about 23.5° to the Equator, it is only at certain times of the day or year that it is displayed to best advantage. It will be realized that the Sun (and therefore the ecliptic) is at its highest in the sky at noon in midsummer, and at its lowest at noon in midwinter. Allowing for the daily motion of the sky, it follows that the ecliptic is highest at midnight in winter, at sunset in the spring, at noon in summer and at sunrise in the autumn. Hence these are the best times to see the planets. Thus, if Venus is an evening object in the western sky after sunset, it will be seen to best advantage if this occurs in the spring, when the ecliptic is high in the sky and slopes down steeply to the horizon. This means that the planet is not only higher in the sky, but will remain for a much longer period above the horizon. For similar reasons, a morning object will be seen at its best on autumn mornings before sunrise, when the ecliptic is high in the east. The outer planets, which can come to opposition (i.e. opposite the Sun), are best seen when opposition occurs in the winter months, when the ecliptic is high in the sky at midnight.

The seasons are reversed in the Southern Hemisphere, spring beginning at the September Equinox, when the Sun crosses the Equator on its way south, summer beginning at the December Solstice, when the

Sun is highest in the southern sky, and so on. Thus, the times when the ecliptic is highest in the sky, and therefore best placed for observing the planets, may be summarized as follows:

	Midnight	Sunrise	Noon	Sunset
Northern latitudes	December	September	June	March
Southern latitudes	June	March	December	September

In addition to the daily rotation of the celestial sphere from east to west, the planets have a motion of their own among the stars. The apparent movement is generally *direct*, i.e. to the east, in the direction of increasing longitude, but for a certain period (which depends on the distance of the planet) this apparent motion is reversed. With the outer planets this *retrograde* motion occurs about the time of opposition. Because of the different inclination of the orbits of these planets, the actual effect is to cause the apparent path to form a loop, or sometimes an S-shaped curve. The same effect is present in the motion of the inferior planets, Mercury and Venus, but it is not so obvious, since it always occurs at the time of inferior conjunction.

The *inferior planets*, Mercury and Venus, move in smaller orbits than that of the Earth, and so are always seen near the Sun. They are most obvious at the times of greatest angular distance from the Sun (greatest elongation), which may reach 28° for Mercury and 47° for Venus. They are seen as evening objects in the western sky after sunset (at eastern elongations) or as morning objects in the eastern sky before sunrise (at western elongations). The succession of phenomena, conjunctions and elongations always follows the same order, but the intervals between them are not equal. Thus, if either planet is moving round the far side of its orbit its motion will be to the east, in the same direction in which the Sun appears to be moving. It therefore takes much longer for the planet to overtake the Sun – that is, to come to superior conjunction – than it does when moving round to inferior conjunction, between Sun and Earth. The intervals given in the table at the top of p.70 are average values; they remain fairly constant in the case of Venus, which travels in an almost circular orbit. In the case of Mercury, however, conditions vary widely because of the great eccentricity and inclination of the planet's orbit.

		Mercury	Venus
Inferior Conjunction	to Elongation West	22 days	72 days
Elongation West	to Superior Conjunction	36 days	220 days
Superior Conjunction	to Elongation East	35 days	220 days
Elongation East	to Inferior Conjunction	22 days	72 days

The greatest brilliancy of Venus always occurs about thirty-six days before or after inferior conjunction. This will be about a month after greatest eastern elongation (as an evening object), or a month before greatest western elongation (as a morning object). No such rule can be given for Mercury, because its distances from the Earth and the Sun can vary over a wide range.

Mercury is not likely to be seen unless a clear horizon is available. It is seldom as much as 10° above the horizon in the twilight sky in northern temperate latitudes, but this figure is often exceeded in the Southern Hemisphere. This favourable condition arises because the maximum elongation of 28° can occur only when the planet is at aphelion (farthest from the Sun), and it then lies well south of the Equator. Northern observers must be content with smaller elongations, which may be as little as 18° at perihelion. In general, it may be said that the most favourable times for seeing Mercury as an evening object will be in spring, some days before greatest eastern elongation; in autumn, it may be seen as a morning object some days after greatest western elongation.

Venus is the brightest of the planets and may be seen on occasions in broad daylight. Like Mercury, it is alternately a morning and an evening object, and it will be highest in the sky when it is a morning object in autumn or an evening object in spring. Venus is to be seen at its best as an evening object in northern latitudes when eastern elongation occurs in June. The planet is then well north of the Sun in the preceding spring months, and is a brilliant object in the evening sky over a long period. In the Southern Hemisphere a November elongation is best. For similar reasons, Venus gives a prolonged display as a morning object in the months following western elongation in October (in northern latitudes) or in June (in the Southern Hemisphere).

The *superior planets*, which travel in orbits larger than that of the Earth, differ from Mercury and Venus in that they can be seen opposite the Sun in the sky. The superior planets are morning objects after conjunction with the Sun, rising earlier each day until they come to opposition. They will then be nearest to the Earth (and therefore at their brightest), and will be on the meridian at midnight, due south in northern latitudes, but due north in the Southern Hemisphere. After opposition they are evening objects, setting earlier each evening until they set in the west with the Sun at the next conjunction. The difference in brightness from one opposition to another is most noticeable in the case of Mars, whose distance from Earth can vary considerably and rapidly. The other superior planets are at such great distances that there is very little change in brightness from one opposition to the next. The effect of altitude is, however, of some importance, for at a December opposition in northern latitudes the planets will be among the stars of Taurus or Gemini, and can then be at an altitude of more than 60° in southern England. At a summer opposition, when the planet is in Sagittarius, it may rise to about 15° above the southern horizon, and so makes a less impressive appearance. In the Southern Hemisphere the reverse conditions apply, a June opposition being the best, with the planet in Sagittarius at an altitude which can reach 80° above the northern horizon for observers in South Africa.

Mars, whose orbit is appreciably eccentric, comes nearest to the Earth at oppositions at the end of August. It may then be brighter even than Jupiter, but rather low in the sky in Aquarius for northern observers, though very well placed for those in southern latitudes. These favourable oppositions occur every fifteen or seventeen years (e.g. in 1988, 2003 and 2018). In the Northern Hemisphere the planet is probably better seen at oppositions in the autumn or winter months, when it is higher in the sky – such as in 2005 when opposition was in early November. Oppositions of Mars occur at an average interval of 780 days, and during this time the planet makes a complete circuit of the sky.

Jupiter is always a bright planet, and comes to opposition a month later each year, having moved, roughly speaking, from one Zodiacal constellation to the next.

Saturn moves much more slowly than Jupiter, and may remain in the same constellation for several years. The brightness of Saturn depends on the aspects of its rings, as well as on the distance from Earth and Sun. The Earth passed through the plane of Saturn's rings in 1995 and 1996, when they appeared edge on; we saw them at maximum opening, and Saturn at its brightest, in 2002. The rings last appeared edge on in 2009.

Uranus and *Neptune* are both visible with binoculars or a small telescope, but you will need a finder chart to help locate them (such as those reproduced in this *Yearbook* on pages 126 and 132). *Pluto* (now officially classified as a 'dwarf planet') is hardly likely to attract the attention of observers without adequate telescopes.

Phases of the Moon in 2010

NICK JAMES

New Moon	d	h	m		First Quarter	d	h	m		Full Moon	d	h	m		Last Quarter	d	h	m
															Jan	7	10	40
Jan	15	07	11		Jan	23	10	53		Jan	30	06	18		Feb	5	23	48
Feb	14	02	51		Feb	22	00	42		Feb	28	16	38		Mar	7	15	42
Mar	15	21	01		Mar	23	11	00		Mar	30	02	25		Apr	6	09	37
Apr	14	12	29		Apr	21	18	20		Apr	28	12	19		May	6	04	15
May	14	01	04		May	20	23	43		May	27	23	07		June	4	22	13
June	12	11	15		June	19	04	30		June	26	11	30		July	4	14	35
July	11	19	41		July	18	10	11		July	26	01	37		Aug	3	04	59
Aug	10	03	08		Aug	16	18	14		Aug	24	17	05		Sept	1	17	22
Sept	8	10	30		Sept	15	05	50		Sept	23	09	17		Oct	1	03	52
Oct	7	18	45		Oct	14	21	27		Oct	23	01	37		Oct	30	12	46
Nov	6	04	52		Nov	13	16	39		Nov	21	17	27		Nov	28	20	36
Dec	5	17	36		Dec	13	13	59		Dec	21	08	14		Dec	28	04	18

All times are UTC (GMT)

Longitudes of the Sun, Moon and Planets in 2010

NICK JAMES

Date		Sun °	Moon °	Venus °	Mars °	Jupiter °	Saturn °	Uranus °	Neptune °
Jan	6	286	177	284	138	327	185	353	325
	21	301	3	303	133	331	185	354	325
Feb	6	317	227	323	127	334	184	354	326
	21	332	50	342	122	338	183	355	326
Mar	6	345	236	358	120	341	183	356	327
	21	0	60	17	121	345	181	357	327
Apr	6	16	282	36	124	348	180	358	328
	21	31	111	55	129	352	179	358	328
May	6	45	313	73	135	355	178	359	329
	21	60	150	91	141	358	178	0	329
June	6	75	357	110	149	0	178	0	329
	21	90	203	128	157	2	178	0	329
July	6	104	30	145	166	3	179	1	328
	21	118	240	162	175	3	180	0	328
Aug	6	133	77	179	184	3	181	0	328
	21	148	287	194	194	2	183	0	327
Sept	6	163	129	208	204	0	185	359	327
	21	178	332	218	214	358	187	359	326
Oct	6	193	167	223	224	356	188	358	326
	21	207	4	220	235	355	190	357	326
Nov	6	223	221	211	246	354	192	357	326
	21	239	50	208	257	353	194	357	326
Dec	6	254	257	213	268	354	195	357	326
	21	269	85	224	280	355	196	357	326

Moon: Longitude of the ascending node: Jan 1: 292°   Dec 31: 272°

Mercury moves so quickly among the stars that it is not possible to indicate its position on the star charts at convenient intervals. The monthly notes must be consulted for the best times at which the planet may be seen.

The positions of the other planets are given in the table on p.74. This gives the apparent longitudes on dates which correspond to those of the star charts, and the position of the planet may at once be found near the ecliptic at the given longitude.

EXAMPLES

In early December 2010, from northern latitudes, two bright planets are seen separated by around 20 degrees in the south-eastern sky about two hours before sunrise. Identify them.

The northern chart 4S (for 6 December at 07h) shows that ecliptic longitudes of 190° to 240° are in the south-eastern sky at the time. Reference to the table of longitudes on page 74 for 6 December shows that Venus (at 213°) and Saturn (at 195°) are the only suitable candidates. Venus is lower down and further east than Saturn – and, of course, it will be by far the brighter of the two.

The positions of the Sun and Moon can be plotted on the star maps in the same manner as for the planets. The average daily motion of the Sun is 1°, and of the Moon 13°. For the Moon an indication of its position relative to the ecliptic may be obtained from a consideration of its longitude relative to that of the ascending node. The latter changes only slowly during the year, as will be seen from the values given on p.74. Let us denote by d the difference in longitude between the Moon and its ascending node. Then if $d = 0°$, $180°$ or $360°$, the Moon is on the ecliptic. If $d = 90°$ the Moon is 5° north of the ecliptic, and if $d = 270°$ the Moon is 5° south of the ecliptic.

On 6 July, the Moon's longitude is given in the table on p.74 as 30° and the longitude of the ascending node is found by interpolation to be about 282°. Thus $d = 108°$ and the Moon is about 5° north of the ecliptic. Its position may be plotted on northern star charts 1S, 9S, 10S, 11S and 12S, and on southern star charts 1N, 9N, 10N, 11N and 12N.

Some Events in 2010

Jan	1	Moon at Perigee (358,680 km)
	3	*Earth* at Perihelion
	4	*Mercury* at Inferior Conjunction
	11	*Venus* at Superior Conjunction
	15	Annular Eclipse of the Sun
	15	New Moon
	17	Moon at Apogee (406,430 km)
	27	*Mercury* at Greatest Western Elongation (25°)
	27	*Mars* nearest to Earth (99.3 million km)
	29	*Mars* at Opposition in Cancer
	30	Full Moon
	30	Moon at Perigee (356,590 km)

Feb	13	Moon at Apogee (406,540 km)
	14	New Moon
	14	*Neptune* in Conjunction with Sun
	27	Moon at Perigee (357,830 km)
	28	*Jupiter* in Conjunction with Sun
	28	Full Moon

Mar	12	Moon at Apogee (406,010 km)
	14	*Mercury* at Superior Conjunction
	15	New Moon
	17	*Uranus* in Conjunction with Sun
	20	Equinox (Spring Equinox in Northern Hemisphere)
	22	*Saturn* at Opposition in Virgo
	28	Moon at Perigee (361,880 km)
	30	Full Moon
	31	*Mars* at Aphelion

Apr	8	*Mercury* at Greatest Eastern Elongation (19°)
	9	Moon at Apogee (405,000 km)
	14	New Moon

24 Moon at Perigee (367,140 km)
28 Full Moon
28 *Mercury* at Inferior Conjunction

May 6 Moon at Apogee (404,230 km)
14 New Moon
20 Moon at Perigee (369,730 km)
26 *Mercury* at Greatest Western Elongation (25°)
27 Full Moon

Jun 3 Moon at Apogee (404,270 km)
12 New Moon
15 Moon at Perigee (365,940 km)
21 Solstice (Summer Solstice in Northern Hemisphere)
26 Full Moon
26 Partial Eclipse of the Moon
26 *Pluto* at Opposition in Sagittarius
28 *Mercury* at Superior Conjunction

Jul 1 Moon at Apogee (405,040 km)
6 *Earth* at Aphelion
11 Total Eclipse of the Sun
11 New Moon
13 Moon at Perigee (361,120 km)
26 Full Moon
29 Moon at Apogee (405,960 km)

Aug 7 *Mercury* at Greatest Eastern Elongation (27°)
10 New Moon
10 Moon at Perigee (357,860 km)
20 *Venus* at Greatest Eastern Elongation (46°)
20 *Neptune* at Opposition in Capricornus
24 Full Moon
25 Moon at Apogee (406,390 km)

Sep 3 *Mercury* at Inferior Conjunction
8 Moon at Perigee (357,190 km)
8 New Moon
19 *Mercury* at Greatest Western Elongation (18°)

Sep 21   Moon at Apogee (406,170 km)
    21   *Jupiter* at Opposition in Pisces
    22   *Uranus* at Opposition in Pisces
    23   Equinox (Autumnal Equinox in Northern Hemisphere)
    23   Full Moon
    23   *Venus* at Greatest Brilliancy (mag. −4.6)

Oct 1   *Saturn* in Conjunction with Sun
     6   Moon at Perigee (359,450 km)
     7   New Moon
    17   *Mercury* at Superior Conjunction
    18   Moon at Apogee (405,430 km)
    23   Full Moon
    29   *Venus* at Inferior Conjunction

Nov 3   Moon at Perigee (364,190 km)
     6   New Moon
    15   Moon at Apogee (404,630 km)
    21   Full Moon
    30   Moon at Perigee (369,440 km)

Dec 1   *Mercury* at Greatest Eastern Elongation (21.5°)
     4   *Venus* at Greatest Brilliancy (mag. −4.7)
     5   New Moon
    13   Moon at Apogee (404,410 km)
    20   *Mercury* at Inferior Conjunction
    21   Full Moon
    21   Total Eclipse of the Moon
    21   Solstice (Winter Solstice in Northern Hemisphere)
    25   Moon at Perigee (368,460 km)

Monthly Notes 2010

January

EARTH is at perihelion (nearest to the Sun) on 3 January at a distance of 147 million kilometres (91.3 million miles).

MERCURY passes through inferior conjunction on 4 January, and then moves rapidly west of the Sun, reaching greatest western elongation (25°) on 27 January. For observers in the latitudes of the British Isles, the planet is inconveniently low in the south-eastern sky before dawn, but for observers in equatorial and southern latitudes it is visible in the early mornings for the last two weeks of the month. It may be seen low above the eastern horizon at the beginning of morning civil twilight. The magnitude of the planet increases from +0.4 to −0.1 during the last two weeks of January.

VENUS passes through superior conjunction on the far side of the Sun on 11 January, when its distance from Earth is 256 million kilometres (159 million miles). Consequently, Venus lies too close to the Sun in the sky to be visible this month.

MARS rises in the early evening and is visible all night long. It has been moving retrograde since 20 December 2009, and it passes from Leo into Cancer during the month. Figure 1 shows the path of Mars against the background stars from January through to the end of August 2010. Mars is at its closest to Earth on 27 January, at a distance of 99.3 million kilometres (61.7 million miles), and reaches opposition just two days later on 29 January, when its magnitude attains −1.3 and the planet's disk has an apparent diameter of just 14.1 seconds of arc. Mars's north pole is tilted towards the Earth.

JUPITER, magnitude −2.1, is visible in the western sky during the early evening. The planet moves from Capricornus into Aquarius early in January.

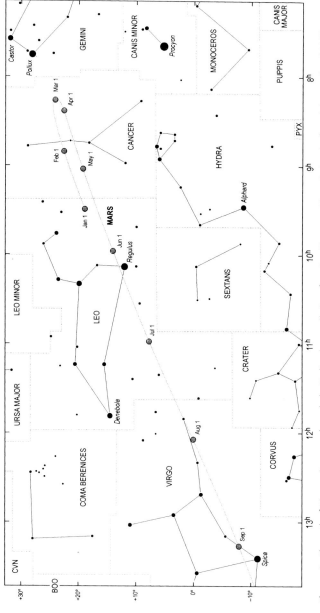

Figure 1. The path of Mars as it moves through the Zodiacal constellations from Leo/Cancer to Virgo during the first eight months of 2010.

SATURN, magnitude +0.8, rises at about midnight. The planet is in Virgo, and after reaching a stationary point on 14 January, its motion is retrograde. Following the ring-plane crossing in September 2009, the rings have been gradually opening again, and are displayed at an angle of 4.8° at the beginning of January, so they should be discernible even in a small telescope.

The Discovery of the Galilean Satellites. It has to be agreed that telescopic astronomy began in 1609, with the work of Thomas Harriot in England. Harriot acquired a telescope and used it to make a map of the Moon which, under the circumstances, was remarkably good. However, Harriot never really followed it up, and the discoveries made by the Italian astronomer Galileo Galilei a few months later had a far greater impact.

Actually Galileo had equipped himself with a telescope in 1609, but at first he concentrated upon its military potential, and it was not until the following January that he turned his attention to the sky. On 7 January 1610, 400 years ago this month, Galileo directed his tiny 'optick tube' towards Jupiter, and saw three star-like objects close to the planet and strung out in a straight line through the planet, parallel to the ecliptic. The next night, the star-like objects had moved from their positions the night before: they were now all located west of Jupiter, and nearer together than on the previous night. So Galileo continued to watch them and Jupiter during the following week. On 13 January, he saw a fourth star-like object. After several weeks, Galileo noted that the four star-like objects never left the neighbourhood of Jupiter and seemed to be carried along with the planet. He also observed that, from night to night, they changed their positions with respect to one other and to Jupiter. Galileo realized that what he was seeing were, in reality, not stars but planetary bodies, i.e. natural satellites, which were in orbit around Jupiter.

Contrary to Church teaching, the Earth was clearly not the only centre of motion in the universe, and the Vatican was at first incredulous, then dismayed. Claims that the telescope was bewitched did not seem very plausible, and it was not long before the satellites were confirmed by other observers. This was really the great 'breakthrough' – a discovery which provided evidence in support of the Copernican system. At least it is encouraging to learn that within the last twenty years the Vatican has not only admitted that Galileo was right, but even

Figure 2. The four Galilean moons of Jupiter, as imaged by the Solid State Imaging system aboard NASA's Galileo spacecraft. Shown from left to right in order of increasing distance from Jupiter, Io is closest, followed by Europa, Ganymede, and Callisto. Io is the most volcanically active body in our Solar System. Europa appears to be strongly differentiated with a rock/iron core, an ice layer at its surface, and the potential for local or global zones of water between these layers. Tectonic resurfacing brightens terrain on the less active and partially differentiated moon Ganymede. Callisto, furthest from Jupiter, appears heavily cratered and shows no evidence of internal activity. (Image courtesy of NASA / JPL / DLR.)

acquitted him of the crime of heresy. The Church has reached this conclusion after a mere 400 years!

Another astronomer of the time, Simon Marius, claimed to have seen Jupiter's satellites as early as late November 1609 (about five weeks before Galileo), but it seems that he did not begin writing down his observations until January 1610 – at about the same time as Galileo was making his first observations. Since Marius did not publish his observations straight away as Galileo had done, it is impossible to establish the truth of Marius' claims. Since Galileo's work was more consistent and extensive, it is he who is nowadays credited with discovering the four largest moons of Jupiter, and they are named the Galilean satellites in honour of their discoverer (Figure 2).

(Incidentally, during his subsequent observations of the moons of Jupiter, on 27 and 28 December 1612, Galileo marked on his sketches the location of what he thought was just an ordinary star. We now know that this 'star' was, in fact, the first known sighting of the distant planet Neptune – made 234 years before its official discovery in 1846.)

Oppositions of Mars. Mars comes to opposition when the Earth, moving around its closer orbit to the Sun, passes between the Sun and Mars. This occurs about every 2 years and 2 months (26 months); the exact figure is the synodic period of 779.94 days. At about these times, the planets also have their closest encounter, but because of their respective orbital eccentricities and orbital inclinations, the exact dates are usually separated by a few days. For example, in 2010, Mars is at its closest to the Earth on 27 January, but doesn't reach opposition until two days later.

This year's opposition of Mars is not particularly favourable insofar as distance is concerned. At the end of January the gap between the two worlds will be 99.3 million kilometres (61.7 million miles). The closest oppositions occur when Mars is also near its perihelion. It must be remembered that while the Earth's orbit is fairly circular, with an eccentricity of only 0.017, that of Mars is much more eccentric at 0.093 (see Figure 3).

This month the apparent diameter of Mars will never exceed 14.1 seconds of arc, and at the next opposition, on 3 March 2012, it will be even smaller: 13.9 seconds of arc, distance 100.8 million kilometres (62.6 million miles). After that things improve: the maximum apparent diameter at the 8 April 2014 opposition will be 15.2 seconds of arc,

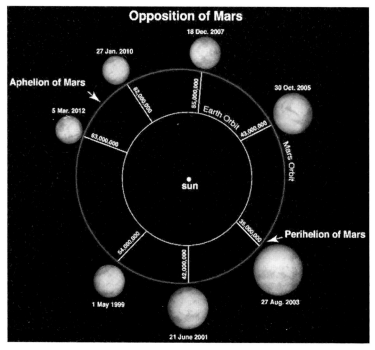

Figure 3. Diagram showing the distance (in miles) of Mars from the Earth at various oppositions. When the Earth passes between the Sun and Mars near Mars' perihelion, as it did in 2003, they are exceptionally close, and Mars appears larger than usual; at other oppositions, the planets are further apart, and Mars appears smaller, as it will in January 2010. (Diagram by Helen Jorjorian / Griffith Observatory.)

then 18.4 seconds of arc on 22 May 2016 and 24.3 seconds of arc at the very favourable opposition of 27 July 2018, when the minimum separation from Earth will be just 57.6 million kilometres (35.8 million miles). Mars doesn't come much closer than this; at the very favourable opposition of Mars in August 2003, the separation distance was 55.8 million kilometres (34.7 million miles).

Yet for Northern Hemisphere observers there is an additional problem. Perihelic oppositions always occur when Mars is well to the south of the Equator, in Sagittarius or Capricornus, and inconveniently low in the southern sky. At least Mars is in Cancer at opposition this year, high up for European and North American observers, so that seeing conditions ought to be good, even though the apparent size of the planet's disc will be rather small.

Brilliant Full Moon. Look at the full moon on the early morning of 30 January. It will look particularly bright, because it is at perigee (its closest point to the Earth) at a distance of 356,590 kilometres (221,580 miles). Of course one full moon and one perigee occur every month, but they do not very often coincide. Incidentally, early on 30 January, the full moon will be in Cancer, just below Mars in the sky, with the Red Planet less than a day past opposition.

February

MERCURY, magnitude −0.1, is visible in the eastern sky before sunrise for observers in the tropics and the Southern Hemisphere during the first two weeks of February, but it eventually becomes lost in the glare of the dawn twilight sky.

VENUS, magnitude −3.9, is slowly drawing out from the Sun and will become visible rather low down in the western sky after sunset from mid-February onwards. On 16–17 February Venus passes only 0.2° south of Jupiter, but for observers in the latitudes of the British Isles this conjunction of the two brilliant planets will be difficult to observe in the bright evening twilight sky. Observers in tropical latitudes, just north of the Equator, will have the best view, although it will be far from easy. The next comparatively favourable conjunction of Venus and Jupiter in the evening sky will be on 15 March 2012.

MARS may be seen in the eastern sky as soon as darkness falls and it is visible for most of the night. The planet is still moving retrograde in Cancer, and by the end of February it lies to the south and slightly east of Pollux (Beta Geminorum). The planet fades from magnitude −1.3 to −0.6 during the month as its distance from the Earth increases.

JUPITER, magnitude −2.0, may be glimpsed low down in the western sky after sunset at the beginning of February. The close conjunction with Venus on 16–17 February has already been mentioned; at this time Venus will appear more than six times brighter than Jupiter, but from the latitudes of the British Isles both objects will be difficult to spot low down in the twilight sky at dusk. Jupiter is in conjunction with the Sun on 28 February, and so will become unobservable towards the end of the month.

SATURN, magnitude +0.7, may be seen rising in the eastern sky in the late evening. The planet is moving retrograde in Virgo, and although the angular width of the rings is about 4°, slightly less than in January, they should still be discernible in all but the smallest instruments.

Orion in the Early Evening. On February evenings, the brilliant winter constellations are on view as soon as the sky gets dark after sunset (Figure 4). Of these, Orion, the Hunter, is pre-eminent; with its characteristic outline, its prominent stars and its surrounding retinue, it dominates the southern aspect of the sky. In mid-February, Orion transits the meridian at about 9.00 p.m, and is visible from dusk until about 3.00 a.m.

Betelgeuse (or Betelgeux), the red M-type giant in the upper left-hand part of the constellation, is a semi-regular variable which varies

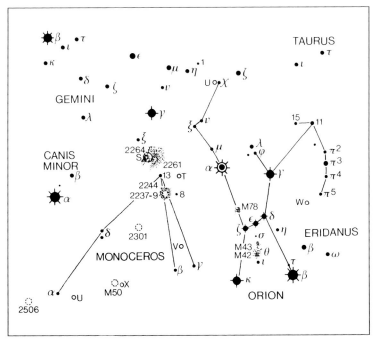

Figure 4. The principal stars of Orion, the Hunter, together with some of the neighbouring constellations. The red giant star Betelgeux (Alpha Orionis) may be found at the upper left of the pattern, while the luminous blue-white star Rigel (Beta Orionis) is located at lower right.

between magnitudes 0.1 and 0.9. Its distance is 430 light years. Betelgeuse is different from the other main stars of Orion, which are generally hot, white and of early spectral type (O or B). Rigel (Beta Orionis), in the Hunter's left foot, which is of spectral type B8, is one such star. Rigel is actually variable over a very small range, but it has a mean apparent magnitude of +0.12, so it is less than one tenth of a magnitude fainter than Capella and Vega, but when seen from the latitudes of Europe and North America it does not appear their equal, simply because it never rises particularly high.

Yet Rigel is vastly more luminous; according to some estimates it is 40,000 times as luminous as the Sun, and lies at a distance of some 775 light years. If the star's invisible ultraviolet radiation is taken into consideration, then the luminosity of Rigel climbs to nearly 70,000 times that of the Sun. Rigel is accompanied by a fairly bright, seventh-magnitude companion, only nine seconds of arc away. Normally such a star would be easily located in a small telescope, but Rigel's intense brilliance almost overwhelms it. With an original mass around seventeen times that of the Sun, Rigel is a dying star, and has most likely reached the stage of fusing internal helium into carbon and oxygen. The star seems fated to explode as a supernova one day.

Luminous though Rigel is, some of the other stars of Orion are comparable. Kappa or Saiph, in the lower left-hand region, has an apparent magnitude of 2.1, so that it is slightly fainter than the Pole Star, but its distance from us is thought to be about 2,100 light years, so that it – like Rigel – is a true 'celestial searchlight'. Gamma, or Bellatrix, and the three stars of the Belt are also highly luminous. And below the Belt may be seen the Hunter's Sword, containing the gaseous nebula Messier 42, a superb sight in a moderate telescope. Altogether, Orion lays claim to being the most spectacular constellation in the entire sky.

The Celestial Crab. When not graced by the presence of a planet, Cancer, the Crab, is one of the least conspicuous of the Zodiacal constellations. There are only two stars above the fourth magnitude, and its brightest star, Beta (Altarf) is only of magnitude 3.5. However, the constellation is easy to identify, as it lies within the large triangle formed by Procyon, Pollux and Regulus. In shape it always reminds me of a very dim and ghostly Orion (Figure 5). However, this month it is host to the Red Planet, Mars, which at magnitude −1.3 at the beginning

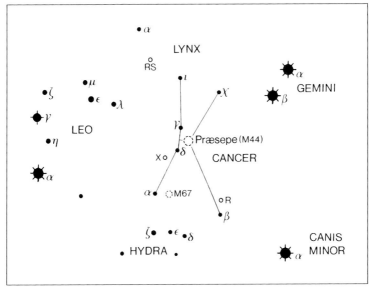

Figure 5. The rather inconspicuous Zodiacal constellation of Cancer, the Crab, is located roughly midway between the stars of Gemini, the Twins, and Leo, the Lion. There are only two stars above the fourth magnitude. The positions of the two open star clusters Praesepe (M44) and M67 are shown.

of February is comparable with Sirius, the brightest star in the sky.

There is a mythological legend attached to Cancer. It represents a crab which the goddess Juno, queen of Olympus, sent to attack Hercules, who was doing battle with the monstrous, multi-headed Hydra. Not unnaturally, Hercules trod on the crab and squashed it! However, as a reward for the crab's efforts, Juno gave it a place in the sky.

Much the most interesting features of Cancer are the two open clusters M44 (Praesepe) and M67. Praesepe is nicknamed the Beehive, though it has also been referred to as the Manger, so that the two stars flanking it, Delta (magnitude 3.9) and Gamma (4.7) are known as the Asses – Asellus Australis and Asellus Borealis respectively. Praesepe was once given a Greek letter, Epsilon, but this has never been used. The cluster is easily visible with the naked eye, but is best viewed with binoculars. It has been known since very early times; around 130 BC Hipparchus referred to it as 'a cloudy star'. It is rather over 570 light years away. There is no nebulosity, and there are many red and orange giant stars, so that Praesepe is clearly older than clusters such as the Pleiades in Taurus. Praesepe is a lovely cluster, so it's hard to see why the Chinese gave it the rather unattractive name of 'The Exhalation of Piled-up Corpses'.

M67, near Acubens or Alpha Cancri (magnitude 4.3), is on the fringe of naked-eye visibility. It is more compact and richer than Praesepe, and again binoculars show it well. It is about 2,700 light years away, and seems to be one of the oldest of all galactic clusters; it lies well over 1,000 light years above the main plane of the galaxy, and so is relatively immune to disruption from field stars. There are at least 500 stars above the sixteenth magnitude.

The variable star X Cancri lies in the same low-power binocular field with Delta. It is a semi-regular star, with a rough period of around 195 days; the spectral type is N, so that the red colour is very pronounced. The range is between magnitudes 5.6 and 7.5, so that the star is always within binocular range. R Cancri, near Beta, is a Mira variable with a range from magnitude 6 to 11.8, and a period of 362 days.

March

Equinox: 20 March

Summer Time in the United Kingdom commences on 28 March.

MERCURY planet passes through superior conjunction on 14 March. Thereafter, the planet moves rapidly east of the Sun, and by the end of the month it is visible in the western sky after sunset for observers in the Northern Hemisphere. Figure 8, given with the notes for April, shows the changes in azimuth and altitude of Mercury at the end of evening civil twilight, about thirty-five minutes after sunset, for observers in latitude 52°N. The changes in the brightness of the planet are indicated on the diagram by the relative sizes of the white circles marking Mercury's position at five-day intervals. The magnitude of the planet decreases from −1.4 to −0.9 during the last week of March.

VENUS, magnitude −3.9, is slowly drawing out from the Sun, and by the end of the month is visible for a short time in the western sky in the early evenings after sunset. The planet's northerly declination means that the period of visibility is rather longer for observers in Europe and North America than it is for those farther south.

MARS is high in the southern sky throughout the evening. The planet reaches its second stationary point on 11 March, and thereafter begins to move direct again through the background stars of Cancer. The planet's magnitude decreases from −0.6 to +0.2 during the month. Mars is at aphelion, its greatest distance from the Sun, on 31 March. On this date it will be 249.2 million kilometres (154.8 million miles) from the Sun, and 150.5 million kilometres (93.5 million miles) from the Earth.

JUPITER was in conjunction with the Sun at the very end of February, and consequently will be unobservable until the last week of March, when it will be seen low down in the eastern sky before dawn for observers in the Southern Hemisphere.

SATURN rises in the early evening and is visible throughout the hours of darkness as it reaches opposition on 22 March, magnitude +0.6. The angular width of the rings has decreased slightly since January to about 3°. The planet is moving retrograde in Virgo, and at opposition its distance is 1,272 million kilometres (790 million miles) from the Earth. Figure 6 shows the path of Saturn against the background stars during the year.

The Globe of Saturn. Saturn reaches opposition this month. It is not as brilliant as it can be – at its best it rises to magnitude −0.3, outshining all the stars apart from Sirius and Canopus – but the rings were edge-wise on last year, and still lie at a narrow angle to us. However, this is a particularly good time to study the globe of Saturn. Our ideas about it have changed markedly during the past century and a half.

Here are the words of R.A. Proctor in 1882: 'Over a region hundreds of thousands of square miles in extent, the glowing surface of the planet must be torn by subplanetary forces. Vast masses of intensely hot vapour must be poured forth from beneath, and rising to enormous heights, must either sweep away the unwrapping mantle of cloud which had concealed the disturbed surface, or must itself form into masses of cloud, recognisable because of its enormous extent.' The 2010 picture is very different. Saturn, like Jupiter, is a gas-giant, and though the core may have a temperature as high as 15,000°C, the upper-most clouds are bitterly cold, at around −180°C. The main constituent of the globe is hydrogen, plus a good deal of helium; there are also hydrogen compounds, notably methane, ammonia and ammonium hydrosulphide; the mean density of the globe is less than that of water. Saturn's globe is so huge that it dwarfs our planet; it would take nine Earths to span its equatorial diameter of 120,536 kilometres (74,900 miles), and it would take 752 globes the size of the Earth to fill its volume. Windspeeds in Saturn's atmosphere may rise to as much as 600 kilometres per hour, and the equatorial rotation period is 10h 14m, faster than for any other planet except Jupiter. Like Jupiter and Neptune, but unlike Uranus, Saturn has a strong internal heat source –

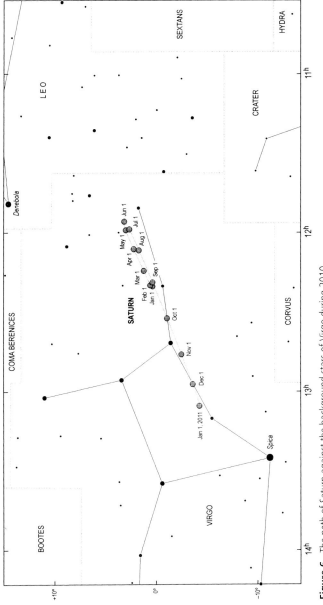

Figure 6. The path of Saturn against the background stars of Virgo during 2010.

thermal energy left over from the planet's birth 4.6 billion years ago. The planet's rocky core is ten to fifteen times as massive as the Earth.

Saturn's magnetic field has a total strength almost 600 times that of the Earth's magnetic field, but only one-thirtieth that of Jupiter's. Saturn's magnetic field is tilted by less than one degree relative to its spin axis, and is reversed in polarity relative to that of the Earth, just like Jupiter's. Electrically charged particles become trapped within the magnetosphere, and these particles spiral inwards at high energies towards the north and south magnetic poles. When these particles hit the upper atmosphere, they excite atoms and molecules there, causing them to glow, producing ovals of auroral light encircling the north and south magnetic poles (Figure 7).

There are darker cloud belts and brighter zones on the globe of the planet, and these are better placed for viewing this year because the rings are relatively unobtrusive. However, a layer of haze above the main cloud belts makes them much harder to discern in amateur telescopes than the cloud belts of Jupiter. Occasional white

Figure 7. This image of Saturn's south polar aurora is an artistic rendition, converting ultraviolet data into the nearest visible colour. Such auroræ are triggered by high-speed particles from the Sun, which electrify the planet's upper atmosphere. The aurora is oval because the glowing gases trace the magnetic field lines that converge, like a narrowing funnel, onto the planet's magnetic poles. (Image courtesy of NASA, ESA, and A. Schaller (for STScI).)

spots may be seen, as in 1933 and 1990, though they do not remain identifiable for long.

This is also a good time to observe the satellites of the planet. Titan (magnitude 8.3) is visible with a very small telescope, and is the only Solar System satellite with a substantial atmosphere; the European Huygens probe, carried to Saturn by NASA's Cassini spacecraft, made a controlled landing there in January 2005 – arguably one of the most difficult space missions successfully completed up to the present time. Rhea, Tethys and Dione are easy telescopic objects; so is Iapetus when west of Saturn. The well-equipped amateur will also be able to see Mimas, Enceladus and Hyperion, but the other members of Saturn's extensive satellite family are much more difficult to see; officially the total number of satellites is now sixty-one.

Make the most of Saturn's unusual appearance this month. The rings will not again be inclined at such a small angle to us until 2025.

The Zodiacal Light. Of the faint, elusive glows in the night sky, the Zodiacal Light is one of the most intriguing. It is never easy to see from the British Isles, but is most likely to be glimpsed in the evenings after sunset in late March/early April, or in the early mornings before sunrise in late September/early October.

Unlike the aurora, the Zodiacal Light originates well beyond the top of the Earth's atmosphere, and seems to be due to the light reflected from a layer of fine interplanetary matter spread out in the main plane of the Solar System. At its best, it may be comparable in brightness to the Milky Way. The Zodiacal Light is best seen from the tropics because from there the ecliptic rises at a steep angle to the horizon at any time of the year. From latitudes north and south of the tropics, it is best seen after sunset in the spring and before sunrise in the autumn.

The Zodiacal Light takes the form of a delicate cone of light, extending along the ecliptic; it is to be seen only when the Sun is below the horizon, though not too far below. When the Sun is only just below the horizon, the sky is too light; when the Sun sinks far enough for the sky to become really dark, the Zodiacal Light usually becomes too faint to be observed, although under ideal conditions, it may be visible throughout the night as a faint band of light following the Zodiac across the sky. Under normal conditions, the best time to view the Zodiacal Light is when the Sun is between 12° and 18° below the horizon, which in late March is between 1h 10m and 2h after sunset in northern Europe.

Even more elusive than the Zodiacal Light is the Gegenschein or Counterglow, which was first described accurately by the Danish astronomer Theodor Brorsen in the nineteenth century. The Gegenschein is a faint, hazy patch of light, always exactly opposite the Sun in the sky, and covering an area which may equal that of the Square of Pegasus. The slightest haze, moonlight, or sky glow from artificial illumination is enough to conceal it, and from the British Isles it is excessively difficult to see. The Gegenschein is really the brightest part of the Zodiacal Band, a dim, parallel-sided extension of the Zodiacal Light. It too must be due to thinly spaced interplanetary material, though its exact origin is still a matter for debate. The Gegenschein is best viewed around local midnight, and from northern temperate latitudes is best seen from the end of December through to the end of January, near the stars of Taurus/Gemini, which are at their highest above the horizon around midnight. During March, the Gegenschein is best viewed from tropical regions north of the Equator (near latitude 10°N), just below the main stars of Leo, at local midnight. Always pick the time of the New Moon when the sky will be at its darkest.

The 'Solitary One'. Look this month for Alphard (Alpha Hydrae), the only bright star in Hydra, the Watersnake, which is the largest constellation in the sky. Its official magnitude is 1.99, and it is easy to locate, partly because of its orange hue and partly because an imaginary line drawn through the Twins, Castor and Pollux, points directly to it. Sir John Herschel regarded it as definitely variable, but this is not easy to check by naked-eye observations, because there are no suitable comparison stars anywhere near it – hence its nickname. It is 177 light years away and 430 times as luminous as the Sun, according to the Hipparcos astrometric satellite; the spectral type is K3.

April

New Moon: 14 April   *Full Moon*: 28 April

MERCURY reaches greatest eastern elongation (19°) on 8 April, and is visible in the western sky in the evenings, for about the first ten days of the month, for observers in tropical and northern latitudes. For Northern Hemisphere observers this is the most favourable evening apparition of the year. Figure 8 shows, for observers in latitude 52°N, the changes in azimuth (true bearing from the north through east, south and west) and altitude of Mercury on successive evenings when the Sun is 6° below the horizon. This is at the end of evening civil

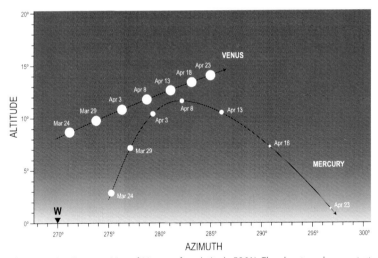

Figure 8. Evening apparition of Mercury, from latitude 52°N. The planet reaches greatest eastern elongation on 8 April. It will be at its brightest in late March, before elongation. The chart also shows the positions of the brilliant Venus, in relation to Mercury, between 24 March and 23 April. Venus is a useful guide to locating the much fainter Mercury between 3 and 8 April. The angular diameters of Mercury and Venus are not drawn to scale.

twilight and in this latitude and at this time of year occurs about thirty-five minutes after sunset.

During its period of visibility, which begins in late March, Mercury fades from magnitude −1.4 to +1.0. The changes in the brightness of the planet are indicated on the diagram by the relative sizes of the white circles marking Mercury's position at five-day intervals: Mercury is at its brightest before it reaches greatest eastern elongation. The diagram gives positions for a time at the end of evening civil twilight on the Greenwich meridian on the stated date. Observers in different longitudes should note that the actual positions of Mercury in azimuth and altitude will differ slightly from those given in the diagram because of the motion of the planet.

The brilliant planet Venus will be in the same part of the sky as Mercury, and will be a useful guide to locating the fainter planet. Between 3 and 8 April, for locations in Europe and North America, at the end of evening civil twilight, Venus and Mercury will be at the same altitude above the horizon, with Venus slightly further towards the west, and about twenty-five times brighter, than Mercury.

Towards the end of the month, Mercury's elongation from the Sun rapidly decreases and the planet passes through inferior conjunction on 28 April.

VENUS is now visible as a brilliant object (magnitude −3.9) in the western sky in the evenings after sunset. Figure 8 also shows, for observers in latitude 52°N, the changes in azimuth and altitude of Venus on evenings in late March and the first three weeks of April when the Sun is 6° below the horizon, i.e. at the end of evening civil twilight. For the first part of April, Venus and Mercury are in the same part of the sky, and the brighter planet may be used as an aid in locating the fainter, more elusive Mercury. On the evening of 16 April, Venus lies below the two-day-old waxing crescent Moon. By the end of April, from the latitudes of the British Isles, Venus sets over two-and-a-half hours after the Sun.

MARS may be found in the southern sky as soon as it gets dark, moving direct in the constellation of Cancer. It fades from magnitude +0.2 to +0.7 during the month.

JUPITER, magnitude −2.0, is still not visible from the latitudes of the British Isles, but for equatorial observers and those in the Southern Hemisphere, the planet may be spotted in the eastern twilight sky before dawn. The planet is in Aquarius, not far from the border with Pisces.

SATURN becomes visible as soon as darkness falls and is observable for most of the night, moving retrograde in Virgo. It fades very slightly from magnitude +0.6 to +0.8 during the month, and the ring angle decreases to only 2° by month end.

Zupi and the Phases of Mercury. Mercury is at greatest eastern elongation on 8 April, but this is not a particularly favourable time to see it, because the elongation is only 19°; at times the elongation may reach almost 28°. Mercury is never a prominent naked-eye object, but it was well known in ancient times. The earliest known recorded observations of Mercury were possibly made by an Assyrian astronomer around the fourteenth century BC. The planet was at first given two names by the ancient Greeks: Apollo when visible in the morning sky and Hermes when visible in the evening. But around the fourth century BC, Greek astronomers realized that the two names referred to the same object. The Romans named the planet after the fleet-footed messenger of the gods, Mercury, whom they equated with the Greek Hermes, because the planet appears to move against the background sky faster than any other planet.

The phases of Mercury are not difficult to see with a telescope of moderate aperture. Although Galileo certainly observed Mercury in the early seventeenth century, and he was able to discern the phases of Venus (see the monthly notes for November), his telescope was not powerful enough to see the phases of Mercury. They were first definitely recorded in 1639 by the Italian astronomer and mathematician Giovanni Battista Zupi (or Zupus), who lived from about 1590 to 1650. Not a great deal is known about him, except that he was born in Catanzaro and was a Jesuit priest. His telescope was only slightly more powerful than Galileo's. He died in Naples. A lunar crater, Zupus, has been named after him.

A Fine Binary Star in Leo. The main spring constellation is Leo, the Lion, in mythology representing the Nemaean Lion who was killed by

Hercules as the first of his Labours, but in the sky Leo is much more imposing than his conqueror. Leo is one of the most prominent of the constellations of the Zodiac, with the distinctive pattern of the Sickle, appearing rather like a backwards question mark, and a separate prominent triangle (Figure 9). The three brightest stars are Alpha (Regulus), magnitude 1.36; Gamma (Algieba), magnitude 1.99; and Beta (Denebola), magnitude 2.14.

Gamma Leonis, or Algieba, is the second brightest star in the Sickle. At magnitude 1.99 it looks much fainter than Regulus and, since it is only very slightly further away from us, Algieba is clearly less luminous. It is also cooler, and even with the naked eye it looks rather orange. Its proper name led the famous nineteenth-century astronomer Admiral W.G. Smyth to remark that the star had been improperly called Algieba, derived from the Arabic *al jeb-bah*, meaning 'the forehead', because no representation of the lion against the background stars of Leo justifies that name. However, proper names are seldom used except for stars of the first magnitude, and a few special cases, so that in general astronomers simply prefer to call the star Gamma Leonis.

It is a beautiful double. The primary is of type K; the companion,

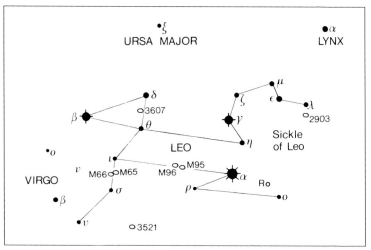

Figure 9. The principal stars of Leo, the Lion, which comprise the distinctive pattern of the 'Sickle' (like a backwards question-mark), and a separate prominent triangle. The fine double star Gamma Leonis, or Algieba, is the second brightest star in the Sickle.

at magnitude 3.5, has a G-type spectrum. Smyth called the primary orange and the companion greenish yellow, but most observers can discern little colour in the fainter star. The separation is over 4 arc seconds, at a position angle of 125°. Their separation is increasing; the pair is a binary system with an orbital period of 619 years, and we are seeing it at a more and more favourable angle. The pair is undoubtedly one of the best doubles in the sky for the user of a small telescope.

A Dim Stellar Neighbour. While in Leo, seek out Wolf 359, RA 10h 56m 29s, Dec +7° 00.9' (J2000), located quite near the ecliptic, some way south and slightly west of the star Theta Leonis, and roughly midway between the stars Sigma and Rho Leonis (see Figure 9). At a distance of 7.8 light years, Wolf 359 is the nearest of our stellar neighbours apart from the three members of the Alpha Centauri system and Barnard's Star. The star was discovered photographically by the German astronomer Max Wolf in 1918. Wolf 359 is a dim red dwarf (just like Proxima Centauri and Barnard's Star), with luminosity a mere 0.00002 of that of the Sun. It is also a flare star with the variable star designation, CN Leonis. Such stars are prone to sudden, unpredictable increases in brightness (flares), before fading back down to normal luminosity within a few minutes. In 1997, the Hubble Space Telescope's Faint Object Spectrograph was used to search for faint companions around Wolf 359. The images obtained revealed no large orbiting bodies (stellar or sub-stellar, such as brown dwarfs) as close as 150 million kilometres (1 astronomical unit) to the star. (The Faint Object Spectrograph was one of the four original axial instruments aboard the Hubble Space Telescope (HST). The instrument was removed from HST during the Second Servicing Mission in February 1997.) You will need a moderate-sized telescope to spot Wolf 359, because its apparent magnitude is only 13.5.

May

New Moon: 14 May   *Full Moon*: 27 May

MERCURY reaches greatest western elongation (25°) on 26 May and towards the end of the month it becomes visible for observers in tropical and southern latitudes. Unfortunately for observers in the latitudes of the British Isles the planet remains unsuitably placed for

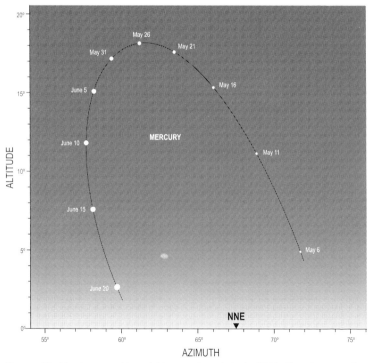

Figure 10. Morning apparition of Mercury, from latitude 35°S. The planet reaches greatest western elongation on 26 May. It will be at its brightest in June, after elongation.

observation throughout the month. For Southern Hemisphere observers this is the most favourable morning apparition of the year. Figure 10 shows, for observers in latitude 35°S, the changes in azimuth (true bearing from the north through east, south and west) and altitude of Mercury on successive mornings when the Sun is 6° below the horizon. This is at the beginning of morning civil twilight and in this latitude and at this time of year occurs about thirty minutes before sunrise.

During its period of visibility, which runs from late May until mid-June, Mercury brightens from magnitude +1.4 to −1.0. The changes in the brightness of the planet are indicated on the diagram by the relative sizes of the white circles marking Mercury's position at five-day intervals. It should be noted that Mercury is at its brightest in mid-June, nearly three weeks after it reaches greatest eastern elongation. The diagram gives positions for a time at the beginning of morning civil twilight on the Greenwich meridian on the stated date. Observers in different longitudes should note that the actual positions of Mercury in azimuth and altitude will differ slightly from those given in the diagram because of the motion of the planet.

VENUS is a brilliant object (magnitude −4.0) in the western sky after sunset. For observers in the latitudes of the British Isles Venus sets nearly three hours after the Sun, but those living further south will experience a somewhat shorter period of visibility. The planet will be occulted by the two-day-old waxing crescent Moon on 16 May – although this event will not be visible from the British Isles – and that evening the Moon and Venus will make a nice pairing in the twilight sky as dusk falls.

MARS continues to be visible in the evening sky, moving direct against the background stars of Cancer towards the border with neighbouring Leo. It continues to decrease in brightness, fading from magnitude +0.7 to +1.0 during the month.

JUPITER magnitude −2.2 is now visible low down in the southeastern sky before dawn for observers in Europe and North America. For those living in equatorial and southern latitudes, the planet rises several hours before the Sun and dominates the eastern sky in the early morning hours. The planet moves from Aquarius into neighbouring Pisces during May.

SATURN is visible as an evening object in Virgo throughout May, setting in the early morning hours. Its brightness decreases from magnitude +0.8 to +1.0 during the month, and the ring angle reaches its minimum for the year, at only 1.7°. On 31 May, Saturn reaches its second stationary point, resuming its direct motion.

A Pioneer in Astronomical Spectroscopy. The English amateur astronomer William Huggins was born in Cornhill, Middlesex, on 7 February 1824, and died a hundred years ago, on 7 May 1910. He was one of the greatest pioneers of stellar spectroscopy, and made a number of fundamental discoveries.

The first real attempt to classify stars according to their spectra was made by Angelo Secchi, in Italy, between 1863 and 1867. He divided stars into four main types:

1. White or bluish stars, with broad, dark hydrogen lines but obscure metallic lines. Example: Sirius. (Remember that to a spectroscopist, all elements heavier than helium are classed as metals, confusing though this seems.)
2. Yellow stars: hydrogen lines less prominent, metallic lines more so. Examples: Capella, the Sun.
3. Orange stars: complicated banded spectra. Examples: Betelgeux, Mira Ceti. This class includes many long-period variable stars.
4. Red stars, with prominent carbon lines; all below apparent magnitude 5. Example: R Cygni. This class also includes many variable stars.

Secchi's four-class scheme prevailed throughout much of the second half of the nineteenth century, and formed the basis for all later classification schemes.

William Huggins enthusiastically followed up Secchi's work. The son of a London silk merchant, and educated privately, Huggins had no financial problems – he was one of the Victorian 'Grand Amateurs' – and he set up an observatory at Upper Tulse Hill in South London, equipping it with a fine 8-inch Alvan Clark refractor. He became fascinated by the spectroscopic work of Gustav Kirchhoff and Robert Bunsen and their studies of the solar spectrum, and he decided that he would try to do the same with the stars. He was helped by his neighbour William Miller, Professor of Chemistry at King's College, London, and in 1875 Huggins married Margaret Lindsay, daughter of a Dublin solicitor, who became his close collaborator and an active astronomer in her own right.

One of Huggins' major discoveries was made in 1864, when he examined the planetary nebula NGC 6543, in Draco, and found that it had an emission or bright-line spectrum. He concluded, correctly, that the nebula was not made of stars, like M31 in Andromeda, but consisted of glowing gas – the first proof of a distinction between starry and gaseous nebulous objects, though it was to be another sixty years before Edwin Hubble showed that M31 and its kind were independent galaxies, lying beyond the Milky Way. You may care to find NGC 6543; RA 17h 58m 33.4s, Dec + 66° 37′ 59.5″ (J2000). The integrated magnitude is 8.8, so that it is not a difficult object. It is not in Messier's catalogue; its nickname is the Cat's-Eye Nebula (Figure 11).

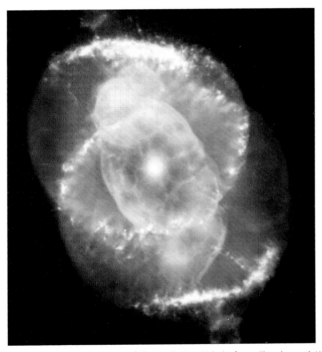

Figure 11. This composite image of the Cat's Eye Nebula from Chandra and Hubble Space Telescope data offers astronomers an opportunity to compare where the hotter, X-ray-emitting gas appears in relation to the cooler material seen in optical wavelengths. (Composite image courtesy of Zoltan G. Levay (Space Telescope Science Institute). The optical images were taken by J.P. Harrington and K.J. Borkowski (University of Maryland) with the Hubble Space Telescope.)

Huggins then examined the spectra of other nebulous objects, and found that many of them, such as M42 in Orion, were gaseous, while others, such as the Andromeda Spiral, were genuinely starry. In 1866, he made the first spectroscopic observation of a nova in outburst, T Coronae Borealis (also nicknamed the 'Blaze Star', and now known to be a recurrent nova), identifying the emission lines of hydrogen. He also obtained the first spectrum of a comet, and found that it contained hydrocarbons. Huggins also extended Secchi's catalogue of stellar spectra.

In 1868, Huggins made another very significant discovery. He examined the spectrum of the brightest star, Sirius, and found a noticeable red shift in its spectrum. It had already been shown by Christian Doppler and Armand Fizeau that the spectral lines of an object moving away from the Earth should be shifted towards the red and those of an object moving towards the observer towards the blue. Since the amount of the shift is proportional to the recession velocity, Huggins was able to work out that Sirius was moving away at about 40 kilometres per second. He soon measured the velocities of many other stars. In 1899, Huggins and his wife published their spectral work in the *Atlas of Representative Stellar Spectra*.

Huggins was knighted in 1897, and was President of the Royal Society between 1900 and 1905. He died at Tulse Hill, and he is buried in Golders Green cemetery; a lunar crater, a Martian crater and an asteroid (2635) are named after him. Lady Huggins survived him by five years.

The Northern Crown. On late evenings in May the constellation of Boötes (the Herdsman) is high up in the south. Boötes is easily recognized because it is dominated by the brilliant golden star Arcturus, which may be found by following round the curve of the handle of the Plough (or the tail of the Great Bear), which at this time of the year is high in the sky, almost overhead. The rest of Boötes also contains some fairly bright stars, including Epsilon (magnitude 2.4), Eta (2.7) and Gamma (3.0). Both Arcturus and Eta Boötis are relatively near neighbours of our Sun, lying at distances of 36 and 32 light years, respectively.

Slightly to the east of Boötes is the little semi-circlet of stars known as Corona Borealis, the Northern Crown. This distinctive constellation is meant to represent the crown given by Bacchus to Ariadne, daughter of King Minos of Crete. This has always been a favourite group with stargazers, and its stars have more pet names and classical allusions than most other patterns. Its brightest star, Alpha or Alphekka, is also known as Gemma and Margarita Coronae. It is 130 times as luminous as our Sun, has an apparent magnitude of 2.2 and lies at a distance of 78 light years. The second brightest star, Beta or Nusakan, at 28 times the Sun's luminosity, is rather closer at 59 light years; its magnitude is 3.7.

In the bowl of the Northern Crown is the star R Coronae Borealis, which is usually just visible to the naked eye, but which at irregular intervals drops rapidly to a minimum so faint that moderate-sized telescopes are needed to see it, and then takes from several weeks to several months to recover. R Coronae variables are deficient in hydrogen but rich in carbon; their unpredictable fadings are due to clouds of 'soot' accumulating in the star's atmosphere.

June

New Moon: 12 June   *Full Moon*: 26 June

Solstice: 21 June

MERCURY remains visible in the eastern morning sky until the middle of the month, for observers in tropical and southern latitudes. Unfortunately for observers in the latitudes of the British Isles the planet remains unsuitably placed for observation throughout the month. For Southern Hemisphere observers this is the most favourable morning apparition of the year. Figure 10, given with the notes for May, shows the changes in azimuth and altitude of Mercury at the beginning of morning civil twilight, about thirty minutes before sunrise, for observers in latitude 35°S. The changes in the brightness of the planet are indicated on the diagram by the relative sizes of the white circles marking Mercury's position at five-day intervals: Mercury is at its brightest in mid-June, when it attains magnitude −1.0. After the middle of June, Mercury draws in towards the Sun and the planet passes through superior conjunction on 28 June.

VENUS, magnitude −4.0, continues to be visible as a brilliant object in the western sky after sunset. The planet's northerly declination, as it moves through Gemini and into Cancer, means that it is far better placed for observers in Europe and North America than it is for those living further south. On the early evening of 15 June, Venus and the three-day-old crescent Moon will present a pleasing spectacle in the western sky as dusk falls. On the same evening, Mars and Saturn will also be visible further east (although very much fainter), the three planets and the Moon strung out in a line clearly marking the arc of the ecliptic; Venus will be near the border between Gemini and Cancer, the Moon will be in Cancer, Mars in Leo and Saturn in Virgo.

MARS is visible in the evening sky, but sets soon after midnight. It has now moved into Leo, and on 7 June, the planet will pass just 0.8°N of

Regulus (Alpha Leonis), Mars being very slightly the brighter of the two. During June, Mars fades from magnitude +1.1 to +1.3.

JUPITER continues to be visible as a conspicuous object in the early morning sky, moving direct amongst the stars of Pisces. Its brightness increases from −2.3 to −2.5 during June. However, observers in equatorial and southern latitudes will have a much better view of the planet than those in Europe and North America. Jupiter passes just 0.4°S of the planet Uranus (magnitude +5.9) on 8 June, and will be a useful guide to locating the fainter planet.

SATURN continues to be visible as an evening object moving direct in Virgo, and still setting after midnight by the end of the month. Its magnitude decreases from +1.0 to +1.1 during June, and the ring angle is only 2.0°.

PLUTO reaches opposition on 25 June, in the constellation of Sagittarius, at a distance of 4,614 million kilometres (2,867 million miles). It is visible only with a moderate-sized telescope, since its magnitude is +14.

The Midnight Sun. The Sun reaches its highest point in the sky from northern temperate latitudes at local noon on 21 June – the summer solstice – when it will be 23½°N of the Celestial Equator. It then rises at its maximum bearing north of due east, and sets north of west, taking more than twelve hours for its passage across the sky, so that we have long days and short nights at this time of year. (From latitude 51°N, there are more than 16½ hours of daylight and from latitude 57°N almost 18 hours of daylight, on 21 June.) From the astronomer's point of view, there is no night at all in these latitudes, for by definition astronomical twilight does not end until the Sun is 18° below the horizon. In the summer months in the British Isles, however, the Sun cannot sink to such an angle below the northern horizon, even at local midnight, so that it never becomes completely dark. Thus in London at the solstice, the Sun is about 15° below the northern horizon at midnight, while in Aberdeen the depression at midnight is only about 9°. The following table shows the dates between which there is no true night at different latitudes:

Latitude

51° (Southern England) from 26 May to 18 July
53° (Midlands) from 16 May to 28 July
55° (Northern England) from 8 May to 5 August
57° (Scotland) from 1 May to 12 August

Still further north, at latitudes greater than 66½°, there will always be at least one day in the year when the Sun does not set, but merely circles the sky during the whole twenty-four hours. This is the so-called 'midnight sun', and a glance at an atlas will show that there are many countries which might claim to be the 'Land of the Midnight Sun' in the Northern Hemisphere. These include the most northerly parts of Norway, Sweden, Finland, Russia, Alaska and Canada, and most of Greenland. In southern latitudes, the midnight Sun will be visible only from Antarctica.

Incidentally, at places lying along the Tropic of Cancer, latitude 23½°N, the Sun will be directly overhead at local noon on 21 June, and at this time a straight pole upright in the ground will lose its shadow.

This Month's Lunar Eclipse. The partial eclipse of the Moon on 26 June will not be particularly impressive. At maximum phase only

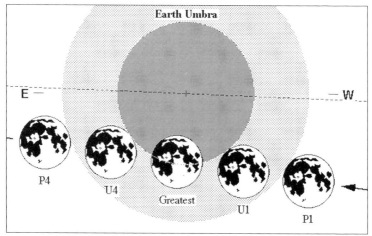

Figure 12. The path of the Moon through the penumbra and umbra during the partial lunar eclipse on 26 June 2010. (Diagram courtesy of Fred Espenak/NASA.)

about 54 per cent of the Moon will be immersed in the southern part of the Earth's umbral shadow (Figure 12). The entire event will be visible from the mid-Pacific Ocean, New Zealand and Australia. From eastern Asia the end of the eclipse may be visible at moonrise, but from Europe and Africa the Moon rises well after the eclipse has ended. Observers in the western USA may get a low-altitude view of the eclipse before moonset. The umbral eclipse lasts from 10h 17m UT to 13h 00m UT; maximum eclipse is at 11h 38m UT.

There are several notable anniversaries this month:

James Short, one of the most prolific telescope makers of the eighteenth century, was born in Edinburgh 200 years ago this month, on 10 June 1710, and was educated at the Royal High School there. His skill in optical work attracted the attention of the Edinburgh University authorities, and before long he was making and selling reflecting telescopes, all of the Gregorian type; his first mirrors were of glass, though he later used metallic specula. His telescopes were of high quality, and some of them can be used even today. He made many observations of comets, the transit of Venus in 1761 and the Northern Lights. He died in London on 15 June 1768.

Wilhelm Heinrich Walter Baade (always known as Walter Baade) was born on 24 March 1893 in Schröttinghausen, Germany, and died fifty years ago, on 25 June 1960. The son of a schoolteacher, Baade was educated at the universities of Münster and Göttingen, and then worked at the Bergedorf Observatory in Hamburg. However, he spent much of his career in the USA, having emigrated there in 1931, but he was always proud of his German nationality, and never took American citizenship.

Together with the highly eccentric Swiss astronomer Fritz Zwicky, Baade carried out important pioneering work on supernovae, and they were the first to propose that supernovae could create neutron stars. Baade wrote several papers with Zwicky, but it must have been an odd partnership: Zwicky referred to Baade as a Nazi, and threatened to shoot him if he found him alone on the campus – a threat which the mild-mannered Baade took very seriously, particularly as Zwicky's appearance was often described as 'threatening'. (He also described his colleagues as 'spherical bastards', because they looked like bastards from whichever direction you observed them.)

But Baade's most important work came later. Taking advantage of the blackout conditions during the Second World War, which reduced light pollution at the Mount Wilson Observatory near Los Angeles, Baade was able to resolve stars in the centre of the Andromeda Galaxy for the first time, using the 100-inch Hooker reflector. This work led him to define two definite 'populations' of stars, known simply as Population I and Population II; Population I were young hot blue stars found in the spiral arms of the Andromeda Galaxy, while Population II stars were older and redder, and were found in the central parts of the galaxy.

Later, Baade discovered that there are two types of Cepheid variable stars. These are short-period variable stars, which vary regularly in brightness and are used to measure distances beyond the Galaxy. Back in 1912, Henrietta Leavitt had discovered that there was a relation-ship between the period of variation of a Cepheid variable and its luminosity (its true intrinsic brightness). In the 1920s, when Edwin Hubble found Cepheids in the outer parts of the Andromeda Galaxy, this period-luminosity relationship had been used to determine the distance to Andromeda as 800,000 light years. Baade showed that the original period-luminosity relationship was true only for Population II Cepheids, whereas Hubble's calculation involved Population I Cep-heids. Baade worked out a new period-luminosity relationship for these new Population I Cepheids, and found that the Andromeda Galaxy was 2 million light years distant. In one short paper, Baade calmly more than doubled the size of the known universe. (I heard that paper when Baade read it at a meeting of the RAS in London. There was a stunned silence when he finished!)

Baade also discovered ten asteroids, including the unusual 944 Hidalgo, which has a long-period orbit, and the Apollo asteroid 1566 Icarus, which can approach the Sun closer than the planet Mercury.

John Kraus, the distinguished American radio astronomer, was born in Ann Arbor, Michigan, a hundred years ago this month, on 28 June 1910. Before and during the Second World War, Kraus was involved in the development of radio antennae, and he also helped construct and operate the University of Michigan cyclotron, then the world's most powerful particle accelerator. In 1946 Kraus began a long-lasting career at Ohio State University, where he rose to become Professor of Electrical Engineering and Astronomy and Director of the Radio

Figure 13. Aerial view of the 'Big Ear' radio telescope which was the brainchild of the distinguished American radio astronomer, John Kraus. The telescope was dismantled in 1998. (Image courtesy of Ohio State University.)

Astronomy Observatory. He invented many novel designs for radio antennae and radio telescopes, which have found widespread use worldwide. Kraus compiled one of the first detailed maps of the radio sky, containing nearly 20,000 radio sources – over half of which had not been previously detected.

Kraus is probably best known for his work with the 'Big Ear' radio telescope (Figure 13), located just south of the city of Delaware from 1963 to 1998. It was part of Ohio State's Search for Extraterrestrial Intelligence (SETI) project. The design of Big Ear, the brainchild of John Kraus, was also used for the Nançay radio telescope in France. Big Ear completed several surveys for extraterrestrial radio sources, but the telescope was dismantled in 1998 to make way for a golf course! John Kraus died on 18 July 2004.

July

EARTH is at aphelion (furthest from the Sun) on 6 July at a distance of 152 million kilometres (94.5 million miles).

MERCURY passed through superior conjunction at the end of June and remains unobservable until the middle of July, when it becomes visible in the western sky after sunset. For observers in the latitudes of the British Isles, the planet is not suitably placed for observation during the month, but for observers in equatorial and southern latitudes this is the most favourable evening apparition of the year. Figure 16, given with the notes for August, shows the changes in azimuth and altitude of Mercury at the end of evening civil twilight, about thirty minutes after sunset, for observers in latitude 35°S. The changes in the brightness of Mercury are indicated on the diagram by the relative sizes of the white circles marking its position at five-day intervals. The planet's brightness decreases from magnitude −0.6 to +0.2 during the last two weeks of July.

VENUS continues to be visible as a magnificent evening object in the western sky after sunset, magnitude −4.1. On 10 July, Venus passes 1° north of Regulus, in Leo, which will be about 140 times fainter than the planet. On 15 July, Venus, Mars, Saturn and the four-day-old crescent Moon form a lovely grouping in the early evening twilight, although Mars and Saturn are rather faint. From the British Isles, about forty-five minutes after sunset, Mars will be located directly above the Moon, while Saturn will be slightly further east. Observers in equatorial and southern latitudes will have a rather better view, and should catch a glimpse of Mercury as well, much lower down in the twilight sky.

MARS, magnitude +1.4, is still visible in the western sky in the evening, but now sets before midnight. For the first half of the month, the planet is in Leo, some way below the three stars making up the triangle at the eastern end of the constellation: Beta (Denebola), Delta and Theta Leonis. Mars moves from Leo into Virgo later in the month and passes 1.8° south of Saturn on 31 July.

JUPITER reaches its first stationary point in Pisces on 24 July, when it begins its retrograde motion. The planet is a brilliant object, brightening from magnitude −2.5 to −2.7 during the month. Figure 14 shows the path of the planet against the background stars during the year.

SATURN continues to be visible in the western sky in the evenings in the constellation of Virgo, magnitude +1.1, although it now sets before midnight.

Intruders into the Zodiac. Conventionally, the Zodiac is divided up into the twelve constellations: Aries, Taurus, Gemini, Cancer, Leo, Virgo, Libra, Scorpius, Sagittarius, Capricornus, Aquarius and Pisces. Since ancient times the Vernal or Spring Equinox has shifted out of Aries into Pisces – and strictly speaking Pisces should now be regarded as the first constellation of the Zodiac instead of the last.

However, there is one notable intruder. This is Ophiuchus, the Serpent Bearer, a large and rather ill-formed constellation which intrudes into the Zodiac for some distance between Scorpius and Sagittarius. Ophiuchus occupies a significant portion of the southern sky during July evenings. The Sun, Moon and planets may therefore move into Ophiuchus; for example, every year, from about 30 November until 17 December, the Sun moves through Ophiuchus, and in June 2007 the planet Jupiter was at opposition in Ophiuchus. And there are other departures, too, from the classic twelve – for example, for some time at the end of June 2005 Mars lay in Cetus, close to the border between Cetus and Pisces.

It is therefore wrong to say that only the twelve famous Zodiacal constellations can be graced with the presence of planets. This fact does not, of course, appear to trouble the astrologers!

There are three notable anniversaries this month:

Gottfried Kirch. The German astronomer Gottfried Kirch died 300 years ago this month. The son of a shoemaker, Kirch was born on

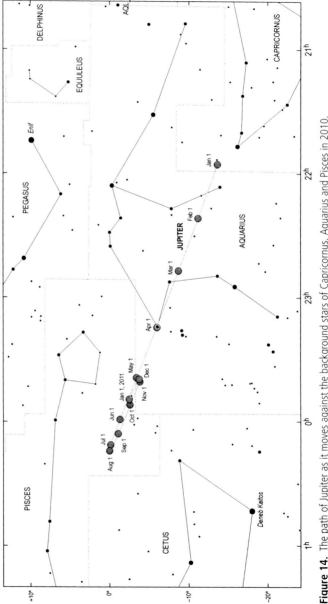

Figure 14. The path of Jupiter as it moves against the background stars of Capricornus, Aquarius and Pisces in 2010.

18 December 1639 in Guben (Saxony). He worked as a calendar-maker, but was able to study astronomy first at Jena and then under Hevelius at Danzig. Subsequently he became a skilled observer, and in 1700 Frederick I of Prussia appointed him First Astronomer of the Prussian Royal Society of Sciences.

Kirch discovered two star clusters, M11 (the Wild Duck) in 1681 and the Serpens globular, M5, in 1702. He was also the first to record the variability of the Mira-type long-period variable star Chi Cygni, in 1686; this was the third variable to be recognized, after Mira itself and Algol (Beta Persei). In his star maps Kirch introduced a new constellation, Sceptrum Brandenburgicum (the Sceptre of Brandenburg) to honour the Brandenburg province of Prussia where he lived. Kirch placed it near the foot of Orion, between Lepus, the Hare, and Eridanus, the River. The pattern did not survive the ruthless twentieth-century pruning by the International Astronomical Union.

Kirch was also an enthusiastic observer of comets, and in 1699 he recorded the comet we now know to have been 55P/Tempel-Tuttle, the parent comet of the Leonid meteor shower seen in November. Kirch died in Berlin on 25 July 1710.

A lunar crater is named after him, and also the asteroid 6841 Gottfriedkirch.

Giovanni Virginio Schiaparelli. This month is the centenary of the death of one of the most famous of all Italian astronomers. Schiaparelli was born on 14 March 1835 in the village of Savigliano, about 50 kilometres south of Turin. He trained as an engineer, graduating from the University of Turin, but he was always keenly interested in astronomy, and worked for a while at the observatories of Berlin and Pulkovo. On his return to Italy, he became Director of the Brera Observatory, near Milan, and remained there until he retired in 1900.

Schiaparelli's most significant contribution to astronomy was probably his discovery that certain periodic comets followed very similar orbits to the meteoroid streams which give rise to annual meteor showers on Earth. In particular he showed that the orbit of the Leonid meteors, seen in November, coincides with that of comet Tempel-Tuttle (now 55P/Tempel-Tuttle), which has a period of about thirty-three years, and that the Perseid meteors, seen every August, have an orbit which coincides with that of comet Swift-Tuttle (now 109P/Swift-Tuttle), which has a period of about 130 years. Incidentally,

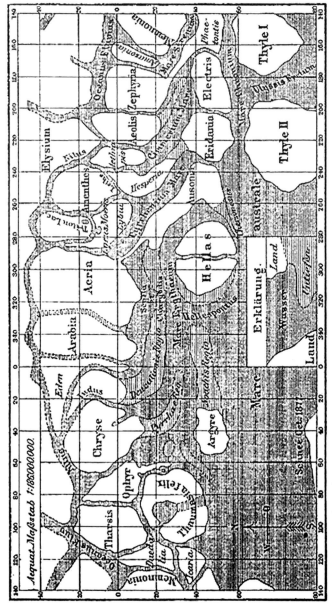

Figure 15. A map of Mars from 1888, drawn by the Italian astronomer Giovanni Schiaparelli, showing his famous *canali* (channels).

the Perseid meteor shower is active from about 23 July until 20 August each year, so expect to see some early members of the shower towards the end of July – although moonlight will be a problem at the end of July 2010.

In 1877, Schiaparelli used the Brera Observatory's excellent 22-centimetre (8.6-inch) refractor to observe Mars, drawing a new map and introducing the nomenclature which, in slightly modified form, is still used today. Schiaparelli also recorded straight, regular lines which he called '*canali*' – channels (Figure 15). Inevitably, this was translated as canals. And equally inevitably there were suggestions that they might represent a network of waterways built by Martian engineers. Schiaparelli kept an open mind about this, but did not regard it as impossible. The canal network does not exist; the 'channels' were mere tricks of the eye.

Though Schiaparelli is best remembered for his observations of Mars, there is no doubt that his most important work was in connection with comets and meteors. He also discovered the asteroid 69 Hesperia on 26 April 1861. Schiaparelli died in Turin on 4 July 1910.

Johann Gottfried Galle. Johann Galle, the distinguished German astronomer, died a hundred years ago, on 10 July 1910. He was born at Pabsthaus, near Wittenberg (Saxony), on 9 June 1812, and went to Berlin University to study astronomy. After graduating, he joined the staff of the Berlin Observatory. It was here that he made the discovery for which he is best remembered.

In Paris, the French astronomer Urbain Le Verrier had studied the perplexing movements of the planet Uranus, and had concluded that the irregularities were due to an unknown planet at a greater distance from the Sun. He worked out a position for it. No immediate search was made from Paris, and so Le Verrier wrote to Galle, asking for help. Galle asked Johann Encke, Director of the Observatory, for permission to use the main telescope – a 23-centimetre (9-inch) Fraunhofer refractor – to see whether he could track down the planet. Encke was reluctant, but finally agreed: 'Let us oblige the gentleman from Paris!' Together with Heinrich D'Arrest, a young member of the Observatory staff, Galle began work, and on the very first night located the planet close to the position given by Le Verrier. After some discussion it was named Neptune.

Some books still give Le Verrier sole credit for the discovery of

Neptune, but this is unfair. The actual discoverers were Johann Galle and Heinrich D'Arrest.

Galle left Berlin in 1851 to become Director of the Breslau Observatory, where he spent the rest of his working life. He was the first to record Saturn's Crêpe Ring, and to measure the parallaxes of several asteroids, notably 8 Flora. He retired at the age of eighty-three, and died in Potsdam at the age of ninety-eight.

August

New Moon: 10 August   *Full Moon*: 24 August

MERCURY reaches its greatest eastern elongation of 27° on 7 August and, for observers in tropical and southern latitudes, is visible as an evening object until the last week of the month. The planet is unobservable for observers in the latitudes of the British Isles. For observers in the Southern Hemisphere this will be the most favourable evening apparition of the year. Figure 16 shows, for observers in latitude 35°S, the changes in azimuth and altitude of Mercury on

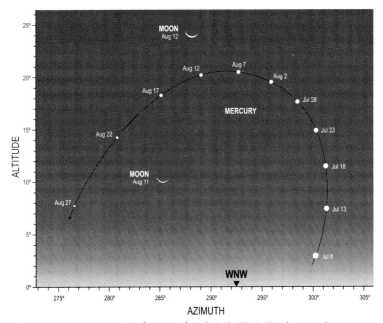

Figure 16. Evening apparition of Mercury, from latitude 35°S. The planet reaches greatest eastern elongation on 7 August. It will be at its brightest in late July, before elongation. The angular diameters of Mercury and the crescent Moon are not drawn to scale.

successive evenings when the Sun is 6° below the horizon. This condition is known as the end of evening civil twilight, and in this latitude and at this time of year occurs about thirty minutes after sunset. The changes in the brightness of the planet are indicated by the relative sizes of the circles marking Mercury's position at five-day intervals: Mercury is at its brightest before it reaches greatest eastern elongation. The diagram gives positions for a time at the end of evening civil twilight on the Greenwich meridian, on the stated date. Observers in different longitudes should note that the observed positions of Mercury in azimuth and altitude will differ slightly from those shown in the diagram. This change will be much greater still for the Moon, if it is shown, as its motion is about 0.5° per hour. During August its brightness fades from magnitude +0.2 to +1.5.

VENUS continues to be visible in the western sky in the early evening after sunset, although for observers in northern temperate latitudes the planet will be inconveniently low in the twilight by mid-August. From the tropics and the Southern Hemisphere it is a magnificent object, brightening from magnitude −4.2 to −4.4 during the month. The planet moves into Virgo at the beginning of August. On the evening of 7 August, Southern Hemisphere observers will see a close grouping of Venus, Mars and Saturn, with Mercury some way below. Figure 17 shows, for observers in latitude 35°S, the azimuths and altitudes of the four planets about thirty minutes after sunset. Venus passes 2.7°S of Saturn on 8 August. The crescent Moon joins the party on 12–13 August. Venus reaches greatest eastern elongation (46°) on 20 August, and on the same day it passes 2.0°S of Mars.

MARS, magnitude +1.5, continues to be visible as an early evening object, in the western sky after sunset, but only for those living in the tropics and the Southern Hemisphere. The planet is in Virgo, and is not far from Venus in the sky during the second half of August.

JUPITER brightens from magnitude −2.7 to −2.9 during the month; it is at opposition next month. The planet is moving retrograde in Pisces. It rises soon after sunset and is then visible throughout the hours of darkness.

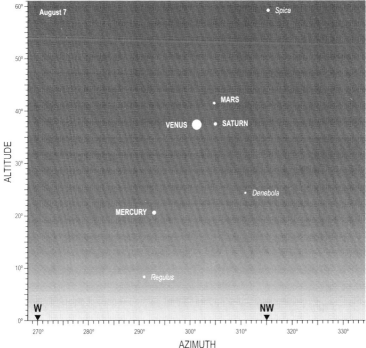

Figure 17. The close grouping of the planets Venus, Mars and Saturn, with Mercury some way below, that will be visible to Southern Hemisphere observers in the evening twilight on 7 August.

SATURN, magnitude +1.0, is also in Virgo. It is no longer visible to observers in the British Isles, but in early August it may be glimpsed rather low down in the western sky after sunset by those living in the tropics and the Southern Hemisphere.

NEPTUNE is at opposition on 20 August, in the constellation of Capricornus. It is not visible with the naked eye since its magnitude is +7.8. At closest approach Neptune is 4,339 million kilometres (2,696 million miles) from the Earth. Figure 18 shows the path of Neptune against the background stars during the year.

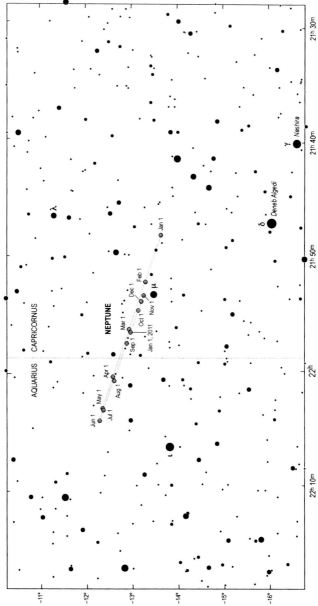

Figure 18. The path of Neptune against the stars of Capricornus and Aquarius during 2010. The moderately bright star near the centre of the chart is Mu Capricorni; the brighter Iota Aquarii is centre left. The stars Deneb Algedi (Delta Capricorni) and Nashira (Gamma Capricorni) are lower right.

Epsilon Aurigae. This month, the attention of astronomers all over the world, both amateur and professional, is focused on one of the most unusual stars in the sky, the eclipsing binary Epsilon Aurigae, which is in the middle of one of its rare eclipses. Its proper name, Almaaz, is seldom used, which seems to be rather a pity.

Epsilon is the star at the apex of a small triangle of stars close to Capella, nicknamed collectively the Haedi (Kids). The others are Eta Aurigae, magnitude 3.2, and Zeta or Sadatoni, which is another very long-period eclipsing binary with a magnitude range between 3.7 and 4.1. It is sheer coincidence that Epsilon and Zeta are so close together because there is absolutely no connection between them; Epsilon is over 2,000 light years away, Zeta only about 800.

Epsilon's primary is an F-type supergiant star, 47,000 times as luminous as the Sun, 18 times as massive and 100 times greater in diameter; its globe could engulf the orbits of all the planets in the Solar System out to and including that of Saturn. Every 27.1 years Epsilon Aurigae begins to fade, and drops to half its normal brightness, as it is eclipsed by a mysterious dark companion. Each eclipse lasts for a total of 714 days – almost two years – with Epsilon taking about six months to fade, remaining faint for about a year, before brightening once more. The present eclipse began in August 2009, reaches its mid-point at the beginning of August this year and ends in May 2011. But what exactly is the nature of the eclipsing secondary? We have to admit that we do not really know.

The changes in brightness were first noticed in 1821 by Johann Fritsch, and confirmed by Eduard Heis and F.W. Argelander, but it was thought that the star was an ordinary irregular variable with a magnitude range of 2.9 to 3.8. In 1904 Hans Ludendorff established that Epsilon is an eclipsing binary, but the secondary component showed no visible light, and there was no sign of a spectrum. Astronomers were baffled. In 1937, the astronomers Gerard Kuiper, Otto Struve and Bengt Strömgren suggested that the secondary was a very large, semi-transparent star, but this did not seem to fit the observations, and neither did the rather wild idea of an eclipsing black hole. Then, in 1965, the Chinese astronomer Su-Shu Huang proposed a large disk system seen almost edge on, and this seems to be the best interpretation so far, probably with a central star; this would explain a slight brightening at mid-eclipse, though it has also been suggested that the dust-disk has a central hole.

During the last eclipse of Epsilon Aurigae in 1981–2, a key portion of the light curve (how the star's apparent brightness varies over time) was lost when the Sun passed below the star in the sky, which it does in early June each year, rendering it unobservable from Earth. Naked-eye estimates may not be precise enough to be really valuable, but they are still worth making. Eta Aurigae (magnitude 3.2) is an excellent comparison star; also use Delta Aurigae (3.7), and 58 Persei (4.3). Do not use Zeta for obvious reasons!

Zeta Aurigae (Sadatoni). Returning to the Haedi, we focus on one of the two stars marking the base of the little triangle – the one at bottom right. This is Zeta Aurigae, another unusual eclipsing binary star. Unlike Epsilon Aurigae, there is no mystery here. It consists of an orange K4-type supergiant star, orbited by a smaller, but hotter, blue B8-type companion. When the blue star passes behind the dimmer, orange companion, Zeta's overall brightness drops by about 25 per cent. Minimum lasts for only 38 days, instead of about a year as with Epsilon. The period is 972 days (2.66 years). At times the light from the B8-star comes to us after passing through the extremely tenuous outer layers of the primary, producing some interesting spectral effects.

September

New Moon: 8 September   *Full Moon:* 23 September

Equinox: 23 September

MERCURY passes through inferior conjunction on 3 September, and rapidly moves out to the west of the Sun, reaching greatest western elongation (18°) on 19 September. The planet is consequently visible from northern and tropical latitudes as a morning object from mid-September until early October. For observers in the northern temperate latitudes this is the best morning apparition of the year. Figure 19 shows, for observers in latitude 52°N, the changes in azimuth and altitude of Mercury on successive mornings when the Sun is

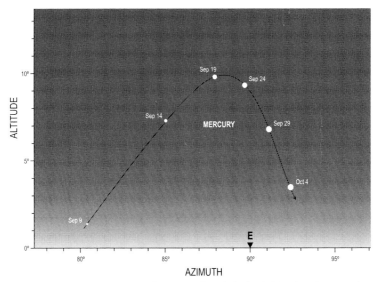

Figure 19. Morning apparition of Mercury, from latitude 52°N. The planet reaches greatest western elongation on 19 September. It will be at its brightest later in the month, after elongation.

6° below the horizon. This condition is known as the beginning of morning civil twilight and in this latitude and at this time of year occurs about thirty-five minutes before sunrise. The changes in the brightness of the planet are indicated by the relative sizes of the circles marking Mercury's position at five-day intervals: Mercury is at its brightest after it reaches greatest western elongation. During its period of visibility its magnitude brightens from +0.5 to −1.1. The diagram gives positions for a time at the beginning of morning civil twilight on the Greenwich meridian, on the stated date. Observers in different longitudes should note that the actual positions of Mercury in azimuth and altitude will differ slightly from those shown in the diagram.

VENUS brightens from magnitude −4.4 to −4.6 during the month, attaining its greatest brilliancy on 23 September. Unfortunately, the planet is now inconveniently low for observers in northern temperate latitudes, but for those in the tropics and the Southern Hemisphere it is a magnificent sight, dominating the western sky after sunset. Venus will be occulted by the three-day-old waxing crescent Moon on 11 September – although this event will not be visible from the British Isles – and that evening observers in the tropics and the Southern Hemisphere will see the Moon and Venus make a nice pairing in the twilight sky as dusk falls.

MARS, magnitude +1.5, continues to be visible low in the western sky after sunset, but only for observers in equatorial and Southern Hemisphere latitudes. The planet may be fairly easily located during the month since it lies not far north of the brilliant Venus. Mars moves from Virgo into Libra during September.

JUPITER, magnitude −2.9, is at its brightest this month since it is at opposition on 21 September and therefore available for observation throughout the night wherever you are on the Earth. The planet is in the constellation of Pisces. Figure 14, given with the notes for July, shows the path of the planet against the background stars during the year. When closest to the Earth its distance is 592 million kilometres (368 million miles).

SATURN is now too close to the Sun for observation, since it passes through superior conjunction at the beginning of October.

URANUS is at opposition on 21 September, in the constellation of Pisces. Near opposition, the planet may be located about 0.8°N of the brilliant Jupiter. Uranus is barely visible to the naked eye as its magnitude is +5.7, but it is easily located in binoculars. Figure 20 shows the path of Uranus against the background stars during the year. At closest approach Uranus is 2,856 million kilometres (1,774 million miles) from the Earth.

The Extreme Tilt of Uranus. Uranus reaches opposition this month, but surface details are by no means easy to make out and even with large instruments the pale greenish disk has a decidedly bland appearance.

For reasons that are still rather unclear, the rotation axis of Uranus is tilted to its orbital plane by 98 degrees – more than a right angle. So, for a planet where the axial tilt is so extreme, how do you decide which is the north pole, and which is the south? The International Astronomical Union (IAU), the controlling body of world astronomy, is quite emphatic about this. All poles above the ecliptic (i.e. the plane of Earth's orbit around the Sun) are north poles, while all poles below the ecliptic are south poles. By this definition, it was Uranus' south pole that was in sunlight during the Voyager 2 spacecraft flyby in January 1986.

The extreme axial tilt also gives the planet a weird calendar. Uranus takes eighty-four years to complete a solar orbit, so each of its four seasons lasts twenty-one years, and near the time of the solstices, one pole is turned continuously towards the Sun, while the other pole is turned away. At such times, near the Equator on the side nearest to the sunward pole, the Sun will be very low down on the horizon. Just over the other side of the Equator it will be dark. Forty-two years later, on the opposite side of Uranus' orbit, the orientation of the poles towards and away from the Sun will be reversed.

Near the time of the equinoxes, the Sun shines down on to the equatorial regions of Uranus, producing more evenly distributed sunlight and giving a period of day and night cycles similar to those seen on other planets. Uranus reached its most recent equinox in December 2007. Consequently, spring (in the northern hemisphere of Uranus) and autumn (in the southern hemisphere) started in December 2007. For the next forty-two years – until 2049 – the north pole will be sunny all of the time, while the south pole will be in darkness.

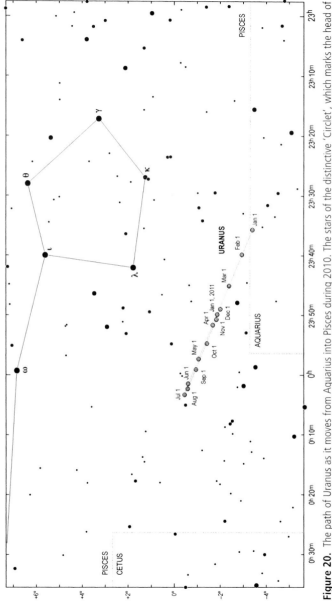

Figure 20. The path of Uranus as it moves from Aquarius into Pisces during 2010. The stars of the distinctive 'Circlet', which marks the head of the western fish in Pisces, and may be found south of the Great Square of Pegasus, are shown towards the top of the chart.

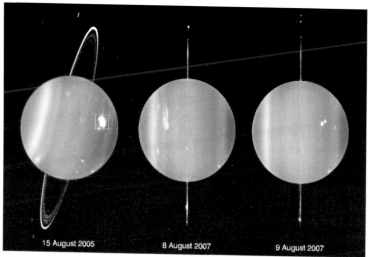

15 August 2005   8 August 2007   9 August 2007

Figure 21. Near-infrared images from the Keck II telescope show the planet Uranus in 2005 (left), with the rings at an angle of 8 degrees, and at equinox in 2007 (right pair), with the planet's ring system edge-on. In all images, the south pole is at the left and the equator is directly below the rings. (Images courtesy of Imke de Pater, University of California, Berkeley; Heidi Hammel, Space Science Institute; Lawrence Sromovsky and Patrick Fry, University of Wisconsin-Madison. Obtained at the Keck Observatory, Kamuela, Hawaii.)

When the Voyager 2 spacecraft flew by in 1986, it was the time of the southern summer solstice, and Uranus appeared virtually featureless. The northern hemisphere of Uranus is now just coming out of the grip of its decades-long winter. As the sunlight reaches some latitudes for the first time in many decades, it will warm the atmosphere and trigger gigantic spring-time storms comparable in size to North America.

In October 2008, at a meeting of the American Astronomical Society's Division for Planetary Sciences, a team led by University of Wisconsin-Madison researcher Lawrence Sromovsky showed crisp new Keck II telescope images of the planet Uranus as it changed seasons (Figure 21). The last time this happened, there were no instruments that could resolve any features on the distant planet, but now it is possible to see what is happening.

Ole Rømer. The great Danish astronomer Ole Christensen Rømer was born in Aarhus, Jutland, on 25 September 1644. He is best remembered for his success in measuring the velocity of light, but this was only one of his many achievements.

Rømer studied astronomy at the University of Copenhagen and was subsequently recruited by Jean Picard, Director of the Paris Observatory, who was visiting Denmark to see Tycho Brahe's observatory at Uraniborg. Rømer joined the Paris Observatory in 1671, where he worked on techniques for the measurement of longitude, and he spent ten years in France, first with Picard and then with the next Director, G.D. Cassini. Rømer did not get on well with Cassini (few people did!) and in 1681 he returned to Denmark, as Professor of Astronomy at Copenhagen University.

Cassini had made observations of the Galilean moons of Jupiter between 1666 and 1668, and Rømer continued this work while in Paris as Cassini's assistant. Adding his own observations to those of Cassini, Rømer found that theory and observation often disagreed: sometimes the eclipses were earlier than predicted, sometimes later. He found that the times between eclipses got shorter as Earth approached Jupiter, and longer as Earth moved farther away. He reasoned that the discrepancies were due to Jupiter's changing distance from Earth: when its distance was greatest, the light had furthest to travel, and the eclipses were late. In 1676 Rømer announced that the speed of light was 225,000 kilometres (140,000 miles) per second. This value was too small but was a good first approximation.

Rømer also invented the transit instrument and the meridian circle. Away from astronomy, he became a highly successful Chief of the Copenhagen Police, and introduced the first street lights (oil lamps). He died on 19 September 1710, 300 years ago this month.

October

New Moon: 7 October   *Full Moon*: 23 October

Summer Time in the United Kingdom ends on 31 October.

MERCURY may be glimpsed in the early morning sky before dawn during the first few days of the month by observers in northern temperate latitudes, but after the first week of October it is lost in the twilight. Figure 19, given with the notes for September, shows the changes in azimuth and altitude of Mercury at the beginning of morning civil twilight, about thirty-five minutes before sunrise, for observers in latitude 52°N. The changes in the brightness of the planet are indicated on the diagram by the relative sizes of the white circles marking Mercury's position at five-day intervals: Mercury is at its brightest in early October, when it attains magnitude −1.2. The planet passes through superior conjunction on the far side of the Sun on 17 October.

VENUS, magnitude −4.6, draws rapidly in towards the Sun after the first week of October and is soon lost in the twilight, even for observers in equatorial and southern latitudes. The planet is not observable after sunset from the British Isles. Venus passes through inferior conjunction on 29 October, and when closest to the Earth its distance is 41 million kilometres (25 million miles).

MARS is unobservable from the British Isles, but for observers in equatorial and southern latitudes, the planet may be spotted low in the western sky at dusk; it lies not far from the brilliant Venus during the first week of October, against the background stars of Libra. The magnitude of Mars is +1.5. Figure 22 shows the path of Mars against the background stars from September through to the end of the year. By the end of October the planet will have become lost in the twilight.

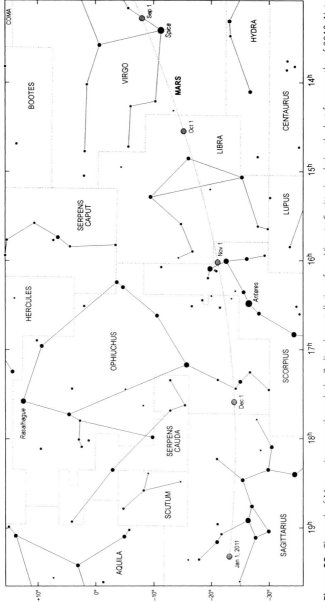

Figure 22. The path of Mars as it moves through the Zodiacal constellations from Virgo to Sagittarius during the last four months of 2010. Note that the planet passes through Ophiuchus, the thirteenth constellation of the Zodiac (see notes for July) between 8 November and 2 December.

JUPITER, magnitude −2.9, is only just past opposition and is visible all night long. The planet's retrograde motion carries it from Pisces back into Aquarius during October. The four Galilean satellites, which Galileo first saw in January 1610, are readily observable with a small telescope or even a good pair of binoculars provided that they are held rigidly.

SATURN passes through superior conjunction on the far side of the Sun on 1 October. Consequently it is unobservable this month.

Tau Ceti. Cetus, the Whale or Sea Monster, is well placed late on October evenings (Figure 23). It is one of the largest constellations in the sky, but not one of the brightest; it contains only two stars above the third magnitude, Beta (Diphda), 2.0, and Alpha (Menkar), 2.4, although the long-period variable star Omicron (Mira) can reach at least 2.0 occasionally. This year, Mira is due to reach maximum brightness on 16 October, so it should be fairly obvious with the naked eye

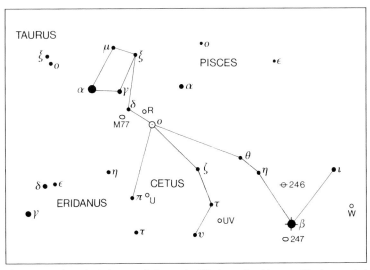

Figure 23. The principal stars of Cetus, the Whale or Sea Monster. The long-period variable star Omicron Ceti (Mira) is shown left of centre. Occasionally, it can brighten to the second magnitude. This year, Mira is due to reach maximum on 16 October, so it should be fairly obvious with the naked eye this month.

this month. There is, however, another star of particular note so far as we are concerned: Tau Ceti, magnitude 3.5. It does have a seldom-used proper name, Durre Menthor.

Tau Ceti is one of our closest stellar neighbours, only twelve light years away. It is of spectral type G8, and it is not too unlike the Sun even though it is only about half as luminous. It is one of only two nearby solar-type stars; the other is Epsilon Eridani. These two were regarded as likely centres of planetary systems, and were carefully studied as soon as suitable equipment became available.

Obviously there was no realistic hope of seeing a planet, but there was at least a chance of radio communication if there were any advanced civilizations in the Tau Ceti system. This led to Project Ozma, organized by Frank Drake in 1960, which involved 'listening out' for any signals rhythmical enough to be interpreted as artificial. Drake had a powerful radio telescope with a diameter of 26 metres (85 feet) at his disposal, working at a frequency of 1.420 gigahertz, sited at Green Bank in West Virginia (Figure 24). But the Tau Cetians and

Figure 24. The 26-metre (85-foot) radio telescope at Green Bank in West Virginia, which was used by astronomer Frank Drake in Project Ozma to 'listen out' for any signals rhythmical enough to be interpreted as artificial. (Image courtesy of SETI League.)

the Epsilon Eridanians remained obstinately silent, and after 200 hours the search was given up. (The project was officially called Ozma, after the famous fictional wizard, though most people referred to it as Project Little Green Men).

There have been different kinds of investigations since then, and with regard to Tau Ceti it has to be said that the results are discouraging. In 2004 a British team led by Jane Greaves discovered that the star is orbited by a vast amount of cometary and asteroidal material – at least ten times more than that of the Sun. Any planet would be mercilessly bombarded, and a team member commented that there would be frequent impacts violent enough to devastate wide areas. Any inhabitants would certainly have to provide themselves with very efficient bomb-proof shelters! The débris disk was discovered by measuring the amount of radiation emitted in the far infra-red portion of the electromagnetic spectrum.

This seems to direct attention away from Tau Ceti and on to Epsilon Eridani, which really does seem to have one planet, probably two. But one can never be sure, and it is always worth having a look at Tau Ceti; you will find it easily enough, close to the slightly fainter Zeta (magnitude 3.7). The two look rather alike, but, as so often in astronomy, appearances are deceptive. Zeta is an orange giant over a hundred times as luminous as the Sun, and about 250 light years away.

Diphda Diphda or Beta Ceti, magnitude 2.0, the brightest star in the Whale, lies south of the Great Square of Pegasus. So does Fomalhaut in Piscis Austrinus (the Southern Fish), southernmost of the first magnitude ever visible from the latitudes of the British Isles. But there is little danger of confusion, because Diphda is almost a magnitude the fainter of the two, and as seen from Britain is much higher above the horizon; locate it by using the two stars which form the left-hand side of the Square of Pegasus (Alpheratz and Algenib) as guides.

November

MERCURY is unobservable at first, having passed through superior conjunction in mid-October, but towards the end of November the planet becomes visible in the evening twilight sky, although only for observers in equatorial and southern latitudes. It will be seen above the west-south-western horizon about the time of end of evening civil twilight. Mercury actually passes 1.7°S of Mars on 20 November, but at magnitude −0.4 Mercury will be six times brighter than the Red Planet at this time.

VENUS passed through inferior conjunction at the end of October, and draws rapidly out from the Sun in November to become visible as a brilliant and spectacular early morning object in the dawn twilight sky. Venus brightens from magnitude −4.1 to −4.7 during the month. The planet, which is in Virgo, is at its best for observers in northern temperate latitudes, from where it rises more than three-and-a-half hours before the Sun by month's end.

MARS is now too close to the Sun for observation, lost in the glare of evening twilight.

JUPITER is visible in the southern sky as soon as darkness falls and does not set until the early morning hours. The planet reaches its second stationary point on 19 November and thereafter resumes its direct motion. The planet is in Aquarius very close to the border with Pisces. It fades from magnitude −2.8 to −2.5 during the month as its distance from the Earth increases.

SATURN passed through superior conjunction in early October and is now becoming visible low in the eastern sky just before dawn. The planet is in Virgo, magnitude +0.9. The rings are now beginning to open again following their edgewise presentation late last year.

Galileo and the Phases of Venus. Venus is on view this month, as a brilliant object in the early morning sky. It was at inferior conjunction on 29 October, and will reach eastern elongation on 8 January next, so that throughout November the phase is steadily increasing.

The crescent stage is interesting because there is always the chance of seeing the Ashen Light, the faint luminosity of the planet's night side. Of course we see the same phenomenon with the Moon ('The Old Moon in the Young Moon's arms' or Earthshine), and there is no mystery about it; it is due to sunlight reflected on to the Moon from the Earth. But Venus has no satellite, and the Ashen Light was not so easy to explain. Attempts were made to dismiss it as a mere contrast effect, but it has been seen by almost every regular observer of the planet, and it can be quite obvious. To test the contrast theory, I once constructed an eyepiece with a curved occulting bar; I positioned the bar so that it blocked out the crescent, and found that the Light was still there. (Try it for yourself. Of course the sky has to be fairly dark, so that Venus will be fairly low down, but this cannot be helped.)

Come now to Galileo. In the autumn of 1610, Venus was visible in the evening sky after sunset, and Galileo made systematic observations of it from late September or the beginning of October right through to the end of the year. By that time he had demonstrated that Venus shows a full cycle of phases, from virtually full to a crescent, but the discovery was not officially announced until New Year's Day 1611. Galileo revealed his discovery by way of an anagram, '*Haec immatura a me iam frustra leguntur o y*', which he had included in a letter to the Tuscan ambassador in Prague. When unscrambled (and translated from the Latin) the anagram said: 'Venus has phases like the Moon.'

The Church was not pleased. With the Ptolemaic theory, with the Earth at the centre of the universe, Venus could not possibly show the full range of phases of this kind. Galileo's January 1610 observations of the satellites of Jupiter had already shown that there was more than one centre of motion in the Solar System, but it was really the behaviour of Venus that proved to be decisive.

(En passant, there are some keen-sighted observers who can make out the crescent form with the naked eye. I certainly cannot – despite my monocle! But the phases are very easy to see with binoculars, particularly when mounted on a tripod.)

Shedir. The familiar 'W' of Cassiopeia is almost overhead during evenings in November. Gamma Cassiopeiae, the middle star of the 'W' (Figure 25), is an unusual eruptive variable star, which has given its name to a category of similar objects. It is usually just below the second magnitude, but it has been known to vary between magnitudes 1.6 and 3.2. Alpha (Shedir) is rated as of magnitude 2.3. I have been making naked-eye estimates of it for over fifty years, and I have always regarded it as slightly variable with a range from 2.1 to 2.5. This would not seem to be surprising because Shedir is a K-type orange giant. However, recent photometric observations have shown no fluctuations at all, so that I am probably wrong. Beta (magnitude 2.37) is a very suitable comparison star.

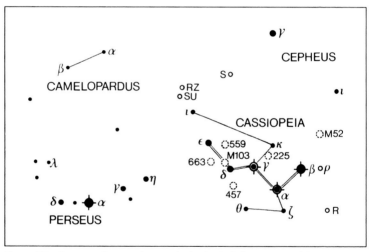

Figure 25. The principal stars making up the familiar 'W' of Cassiopeia. Alpha Cassiopeiae (Shedir) is an orange K-type giant star that has long been suspected of variability. Its magnitude is given as 2.3. In mythology, Cassiopeia was the proud and boastful queen who was mother to the beautiful Andromeda and wife of Cepheus.

December

New Moon: 5 December   *Full Moon*: 21 December

Solstice: 21 December

MERCURY reaches greatest eastern elongation (21°) on 1 December and is visible as an evening object to observers in tropical and southern latitudes during the first week of the month, when the planet can be seen in the west-south-western sky about the time of end of evening civil twilight. Mercury fades rapidly from magnitude -0.4 to $+0.5$ during the first ten days of December. The planet will lie 1.7°S of the two-day-old crescent Moon on 7 December. Mercury passes through inferior conjunction on 20 December and becomes visible again, low in the eastern sky before dawn by month end, again for observers in equatorial and more southerly latitudes.

VENUS attains its greatest brilliancy at magnitude -4.7 on 4 December. It is a magnificent object in the early morning sky before dawn, rising more than four hours before the Sun by mid-December for observers in the latitudes of the British Isles. On the early morning of 2 December the waning crescent Moon will make a nice pairing with Venus in the dawn twilight sky. The much fainter Saturn will lie a little way above the pair at this time.

MARS remains too close to the Sun for observation this month. The planet will pass through superior conjunction on the far side of the Sun in February 2011, and there will be no opposition of Mars in 2011.

JUPITER is a splendid early evening object, visible in the southern sky as soon as it gets dark and setting around midnight. The planet is moving direct once more and its motion carries it from Aquarius back into Pisces during December. The planet fades slightly from magnitude -2.5 to -2.3 in the course of the month.

SATURN, magnitude +0.8, is moving eastwards in Virgo and it continues to be visible in the eastern sky before dawn.

Sputnik 6. Almost everyone remembers Sputnik 1, which ushered in the Space Age in October 1957, but who now remembers Sputnik 6, also known as Korabl-Sputnik 3 (K-S 3), which was launched fifty years ago this month – on 1 December 1960? Very little was heard about it at the time because the Cold War had started, and not much information came out of the USSR. We were told, however, that this was a test flight for the Vostok missions. K-S 3 carried no astronauts; the passengers were two dogs, Pchelka ('Bee') and Mushka ('Little Fly'), mice, insects and plants.

The spacecraft carried a television camera and an assortment of scientific instruments, all of which apparently worked well. The flight lasted for only a day, and on its eighteenth orbit of the Earth, K-S 3 fell back into the atmosphere and burned up during too steep a re-entry. In any case, it carried an explosive charge to prevent any chance that it would land intact and fall into foreign hands.

There was no plan to make a safe controlled landing, and so no hope that Pchelka and Mushka would survive. No doubt their flight paved the way for Yuri Gagarin's mission in the following year, but some people voiced strong objections to the use of dogs and other intelligent animals in this way. I was totally against it, but I can see why the Soviets did it.

Eclipse at the Winter Solstice. On 21 December there will be a total eclipse of the Moon. Totality begins at 07h 40m UT, reaches maximum at 08h 17m UT, and ends at 08h 54m UT. The entire eclipse will be visible after local midnight from Canada, the USA, central America and the north-western tip of South America. From southern parts of the British Isles, the initial umbral phases will be visible, but the Moon will be dropping down into the western sky as dawn approaches. The eclipse first becomes total at 07h 40m UT with the Moon very low in the west-north-western sky, close to the horizon and in a rapidly brightening sky. From London, sunrise is at 08h 04m UT and moonset just eight minutes later. The umbral phase lasts from 06h 32m UT until 10h 02m UT. During totality, the Moon tracks through the northern part of the Earth's umbral shadow (Figure 26), so for those observers

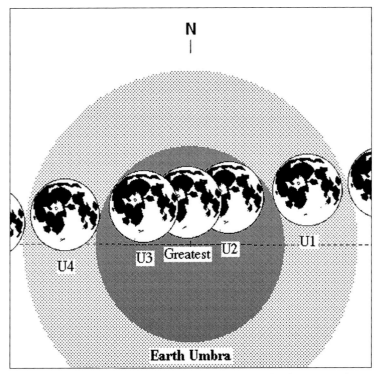

Figure 26. The path of the Moon through the penumbra and umbra during the total lunar eclipse on 21 December 2010. (Diagram courtesy of Fred Espenak / NASA.)

watching it high up in a clear sky, the southern half of the totally eclipsed Moon should appear considerably darker than the northern part.

One never quite knows how dark or how bright a lunar eclipse will be. Everything depends on the conditions in the Earth's upper atmosphere through which all light falling on to the shadowed Moon has to pass. I have seen eclipses when the Moon has been difficult to find even with a telescope, while at other eclipses it has remained bright red or vividly coloured. A lunar eclipse may lack the glory of a total eclipse of the Sun, but it has a quiet beauty all its own.

There will be two total lunar eclipses in 2011, on 15 June and 10 December.

The Demon Star. The constellation Perseus is almost overhead from Europe and North America during December evenings (Figure 27). In mythology Perseus was the Greek hero who wore winged shoes, loaned to him by Mercury, messenger of the gods. He was sent on a mission to cut off the head of the hideous Gorgon Medusa, a fearsome woman with snakes for hair, who could turn any living creature to stone with just one look. Perseus chopped off her head by looking at her reflection in his shield, but on the way back from this adventure he rescued the beautiful Andromeda, daughter of Cassiopeia and Cepheus, who was chained to a rock, to be sacrificed to a sea monster (probably represented by the pattern of Cetus in the night sky). In the best traditions of such legends, Perseus married Andromeda and they all lived happily ever after!

The brightest star in Perseus is Alpha Persei or Mirphak, which

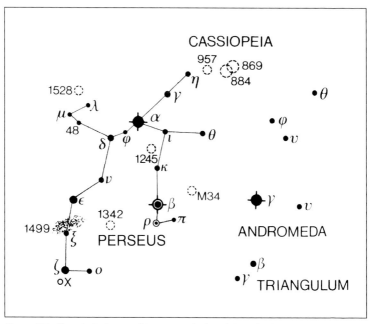

Figure 27. The principal stars of Perseus, with the celebrated eclipsing binary star Algol (Beta Persei), also known as the 'Demon Star' or 'Winking Demon', shown below and slightly left of centre. Every 2½ days, the star fades by over a magnitude, when the larger, dimmer orange star in the system eclipses its brighter blue companion.

is slightly above the second magnitude, but the most famous object in the pattern is Beta, better known by its proper name of Algol, one of the most celebrated variable stars in the sky. Algol is also known as the 'Demon Star' or the 'Winking Demon', because it is said to mark the position of Medusa's evil eye; its variations in brightness were certainly noted in ancient times.

Usually Algol shines at magnitude 2.1, and remains constant (or nearly so) for two-and-a-half days. It then starts to fade, taking five hours to drop down to magnitude 3.3; after a brief minimum, it takes another five hours to regain its normal brightness. Behaviour of this sort is quite different from that of an ordinary variable, and in fact Algol is not intrinsically variable at all. It is an eclipsing binary consisting of a bright, hot blue star, and a slightly larger orange star which is much dimmer than its companion. When the blue star passes in front of the orange star the total light from the two stars decreases only slightly. But when the larger, dimmer orange star passes in front of the brighter blue star, the total light decreases significantly and the star appears to 'wink'. The fluctuations are easily visible with the naked eye, and the times of minima are given in the *Handbook of the British Astronomical Association*.

Many Algol-type variables are now known. The second brightest member of the class is Lambda Tauri, where the magnitude range is from 3.3 to 4.2, and the period 3.9 days.

Eclipses in 2010

MARTIN MOBBERLEY

During 2010 there will be four eclipses, two of the Sun and two of the Moon.

1. *An annular eclipse of the Sun* on 15 January will be visible as a partial eclipse from Africa, the Indian Ocean and Asia, but will be invisible from most of Europe (except from a few Mediterranean locations at sunrise), Australia and the USA. The partial phase begins (in Africa) at 04h 05m UT and ends (in China) at 10h 08m UT. The narrow path of annularity starts in central Africa, crosses the Indian Ocean, clips the Maldives, the southern tip of India and northern Sri Lanka, and then hits land at Burma. The final part of the annular track is across China

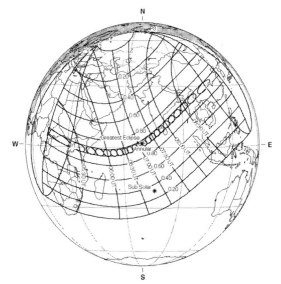

Figure 1. The track of the 15 January 2010 annular solar eclipse. (Diagram courtesy of Fred Espenak / NASA.)

(Figure 1). Cities experiencing annularity include Kampala and Nairobi in Africa, Mandalay in Burma, and Xiaguan and Chongquing in China. Annularity begins at 05h 14m UT and ends at 08h 59m UT. The maximum duration of annularity is 11 minutes 8 seconds in the Indian Ocean, with a track width of 335 kilometres (208 miles). Because the Sun appears so much larger than the Moon at this eclipse, the maximum duration of the annular phase – 11 minutes 8 seconds – will not be exceeded until the year 3043!

2. *A partial eclipse of the Moon* on 26 June is visible in its entirety from the mid-Pacific Ocean, New Zealand and Australia (from the extreme west of Australia the Moon rises just after the start of the penumbral phase). From eastern Asia the latter stages of the eclipse may be visible at moonrise but the Moon will rise well after the eclipse has ended from Europe and Africa. Observers in the western United States may get a low-altitude view of the eclipse (cloud permitting) before the Moon sets. At the maximum of the eclipse (11h 38m UT) only 54 per cent of the Moon will be in the umbra; specifically, the northern lunar hemisphere will be immersed in the southern part of the umbra. The Moon first touches the umbra at 10h 17m UT and leaves at 13h 00m UT.

3. *A total eclipse of the Sun* on 11 July is visible as a partial eclipse from the south Pacific Ocean and, at sunset, from the countries of southern South America. The partial phase begins at 17h 10m UT and ends at 21h 57m UT. The path of totality starts over 1,500 kilometres (1,000 miles) north-east of New Zealand, tracks north-north-east for over 3,000 kilometres (2,000 miles) through French Polynesia, and then curves south-east, crossing Easter Island and its famous stone statues, before hitting the coast of southern Chile and ending at sunset in southern Argentina (Figure 2). Therefore, excluding cruise ships, the only land-based locations for viewing this eclipse are the French Polynesian islands such as Tupapati or Ikitake, Easter Island itself or the western coast of the extreme southern tip of Chile (latitude 49° S). In practice the Sun will be too low for useful totality observations from Argentina. Totality begins at sunrise at 18h 51m and ends at sunset at 20h 52m. The maximum duration of totality on the centre line is 5 minutes 20 seconds in the south Pacific Ocean, where the track will be 257 kilometres (160 miles) wide.

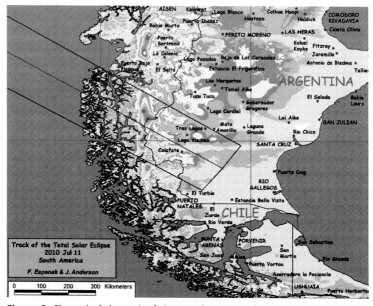

Figure 2. The end of the track of the 11 July 2010 total solar eclipse in Chile and Argentina. (Diagram courtesy of Jay Anderson / Fred Espenak.)

4. *A total eclipse of the Moon* on 21 December is visible in its entirety from Canada, the USA, central America and the north-western tip of South America. From southern parts of the UK, the initial umbral phases will be visible with the Moon low in the south-western sky as dawn approaches, but the eclipse first becomes total at 07h 40m UT with the Moon setting, close to the UK horizon and in a bright, pre-sunrise sky. From Australia and eastern Asia the reverse is true and the totality phase is just observable near moonrise from eastern Australia, north-eastern China and Japan; further west the totality phase will be invisible. The entire eclipse is completely invisible from southern and eastern Africa, the Middle East and India. At maximum eclipse (08h 17m UT) the Moon sits within the northern part of the umbra, so this total lunar eclipse should see the Moon with a relatively bright northern limb and a much darker southern limb. Totality ends at 08h 54m UT. The Moon enters and leaves the umbra at 06h 32m and 10h 02m UT respectively.

Occultations in 2010

NICK JAMES

The Moon makes one circuit around the Earth in just over twenty-seven days and as it moves across the sky it can temporarily hide, or occult, objects that are further away, such as planets or stars. The Moon's orbit is inclined to the ecliptic by around 5.1° and its path with respect to the background stars is defined by the longitude at which it crosses the ecliptic passing from south to north. This is known as the longitude of the ascending node. After passing the node the Moon moves eastward relative to the stars, reaching 5.1° north of the ecliptic after a week. Two weeks after the ascending node it crosses the ecliptic moving south (the descending node), and then it reaches 5.1° south of the ecliptic after three weeks. Finally it arrives back at the ascending node a week later and the cycle begins again.

The apparent diameter of the Moon depends on its distance from the Earth, but at its closest it appears almost 0.6° across. In addition, the apparent position of the Moon on the sky at any given time shifts depending on where you are on the surface of the Earth. This effect, called parallax, can move the apparent position of the Moon by just over 1°. The combined effect of parallax and the apparent diameter of the Moon means that if an object passes within 1.3° of the apparent centre of the Moon as seen from the centre of the Earth it will be occulted from somewhere on the surface of our planet. For the occultation to be visible the Moon would have to be some distance from the Sun in the sky and, depending on the object being occulted, it would have to be twilight or dark.

For various reasons, mainly the Earth's equatorial bulge, the nodes of the Moon's orbit move westwards at a rate of around 19° per year, taking 18.6 years to do a full circuit. This means that, while the Moon follows approximately the same path from month to month, this path gradually shifts with time. Over the full 18.6-year period all of the stars that lie within 6.4° of the ecliptic will be occulted.

Only four first-magnitude stars lie within 6.4° of the ecliptic. These are Aldebaran (5.4°), Regulus (0.5°), Spica (2.1°) and Antares (4.6°). As the nodes precess through the 18.6-year cycle there will be a monthly series of occultations of each star followed by a period when the star is not occulted. In 2010 the only first-magnitude star to be occulted will be Antares.

In 2010 there will also be four occultations of bright planets: three of Venus and one of Mars. Only two of these events take place at a solar elongation of greater than 20° and neither is visible from the British Isles.

The following table shows events potentially visible from somewhere on the Earth when the solar elongation exceeds 20°. More detailed predictions for your location can often be found in magazines or the *Handbook of the British Astronomical Association*.

Object	Time of Minimum Distance (UT)	Minimum Distance °	Elong- ation °	Visibility
Antares	11 Jan 2010 13:10	1.1	−41	Northern Canada
Antares	7 Feb 2010 18:56	1.1	−69	Northern Pacific
Venus	16 May 2010 10:18	0.1	30	India, Middle East, North Africa
Venus	11 Sep 2010 12:56	0.3	43	Southern Africa

Comets in 2010

MARTIN MOBBERLEY

Some forty short-period comets will reach perihelion in 2010. All these imminently returning comets orbit the Sun with periods of between three and twenty years and many are too faint for amateur visual observation, even with a large telescope. Bright, or spectacular, comets have much longer orbital periods and, apart from a few notable exceptions such as 1P/Halley, 109P/Swift-Tuttle and 153P/Ikeya-Zhang, the best performers usually have orbital periods of many thousands of years and are often discovered less than a year before they come within amateur range. For this reason it is important to regularly check the best comet websites for news of bright comets that may be discovered well after this *Yearbook* is published. Some recommended sites are:

British Astronomical Association Comet Section: www.ast.cam.ac.uk/~jds/
Seiichi Yoshida's bright comet page: www.aerith.net/comet/weekly/current.html
CBAT/MPC comets site: www.cfa.harvard.edu/iau/Ephemerides/Comets/index.html
Yahoo Comet Images group: http://tech.groups.yahoo.com/group/Comet-Images/

The CBAT/MPC web page above also gives accurate ephemerides of the comets' positions in right ascension (RA) and declination (Dec.).

Six periodic comets are confidently expected to reach perihelion at a magnitude of 12 or brighter during 2010, while still being observable in a reasonably dark sky. A seventh periodic comet, 157P/Tritton, may possibly outburst to magnitude 12 in February if it performs as in 2003. One non-periodic comet, 2007 Q3 (Siding Spring), may still be magnitude 11 in early 2010 despite reaching perihelion the previous October. One further periodic comet, close to aphelion, may also be of interest, especially to CCD observers. Comet 29P/Schwassmann-Wachmann orbits the Sun every 14.6 years and can be as faint as magnitude 19. However, it goes into outburst frequently, sometimes peaking at magnitude 11 or even 10. It will probably sit around magnitude 15 for much of 2010 and will be best placed in mid-February not far from Regulus in Leo.

These comets are listed below in the order in which they are expected to peak.

Comet	Period (years)	Perihelion	Peak Brightness
C/2007 Q3 (Siding Spring)	long	2009 Oct 7.3	10 in prev. year
157P/ Tritton	6.30	2010 Feb 20.5	12 max. in Feb
118P/Shoemaker-Levy	6.45	2010 Jan 2.3	12 in Jan
29P/Schwassmann-Wachmann	14.65	2004 Jun 29.4	11 in outburst
81P/Wild	6.42	2010 Feb 22.7	9 in Feb/Mar
65P/Gunn	6.79	2010 Mar 2.1	12 in Mar
10P/Tempel	5.37	2010 Jul 4.9	10 in Jul
2P/Encke	3.30	2010 Aug 6.5	7 in Aug
103P/Hartley	6.47	2010 Oct 28.3	4 or 5 in Oct/Nov

As 2010 starts, three short-period comets, 157P/Tritton, 118P/Shoemaker-Levy and 81P/Wild, may all be magnitude 12 or brighter and the more recently discovered long-period comet 2007 Q3 Siding Spring should still be magnitude 11 despite peaking in 2009. 118P will be very well placed in January quite close to Betelgeuse in Orion, but at magnitude 12 at best it will not be a show-stopping sight. Whether 157P actually makes an appearance largely depends on whether it is in outburst. During January it will be an evening object close to Algenib in Pegasus, but it may well be so faint that only a CCD will reveal it. The only long-period comet in this list, C/2007 Q3 (Siding Spring), rises after 22.00 hours in January and is best after midnight, not far from the globular cluster M3 on the Boötes/Canes Venatici border. The only really promising comet in early 2010 is 81P/Wild. It could be magnitude 10 in January, but as it will be close to Saturn in Virgo, you will not get a good view of it until after midnight in those bone-chilling UK pre-dawn winter skies. Comet 81P/Wild reaches perihelion on 22 February and will be best in February and March, when it should peak at a healthy magnitude 9, while still in Virgo (Figure 1). Some bits of comet Wild live in a laboratory here on Earth! The Stardust spacecraft visited the comet in 2004 and successfully crash-landed the samples back here two years later.

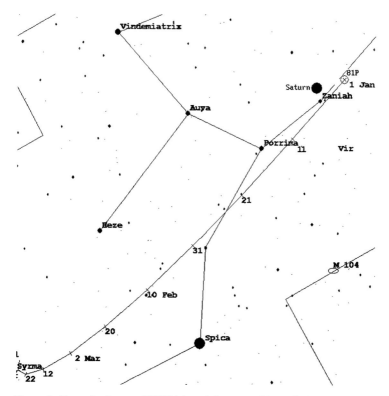

Figure 1. The path of comet 81P/Wild from 1 January to 22 March 2010 as it moves against the background stars of Virgo. The comet passes just south of the planet Saturn in early January. Comet 81P/Wild reaches perihelion on 22 February and will be at its best in February and March when it should peak at a magnitude 9, while still in Virgo.

As mentioned earlier, CCD imagers may well benefit from monitoring the position of comet 29P/Schwassmann-Wachmann in Leo during the early months of the year, as it can go into outburst at any time, even if it can sink below magnitude 15 for long periods.

The next comet worthy of serious amateur attention this year will be 65P/Gunn, which will be reasonably placed for Southern Hemisphere observers in the pre-dawn March skies, where it will track through eastern Sagittarius. However, it is not expected to be brighter than magnitude 12, so may primarily be a CCD target. In the late Northern Hemisphere summer months, comet 10P/Tempel struggles to escape from the long UK summer twilight in July, but it is not an impossible object if you have a good south-east dawn horizon. 10P/Tempel should reach magnitude 10 or even 9, but is better placed for Southern Hemisphere observers.

Comet 2P/Encke is our most regular cometary visitor, with a 3.3-year orbit, but it only becomes really bright when it is very close to the Sun and therefore can be a real challenge to observe. This year it will be best placed for Southern Hemisphere observers after perihelion – that is, from late August onwards, when it should be magnitude 8 or 9 as it speeds through southern Virgo, clipping north-east Corvus and

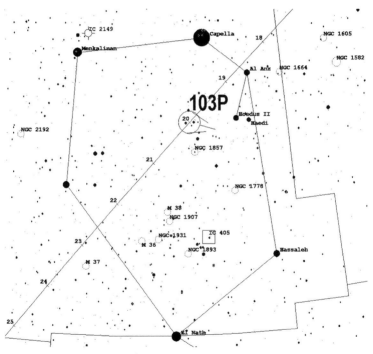

Figure 2. The expected path of Comet 103P/Hartley as it plunges south through Auriga between 18 and 25 October 2010. The comet should pass only 0.12 AU from the Earth on 20 October when it may be 4th or 5th magnitude. Sadly, the Moon will be only three days from full at this time, which will make naked eye sightings difficult. As the comet is making such a close flyby of Earth even a slight change in the date of perihelion will change the comet's path in October quite noticeably.

passing close to Messier 104 on 30 August. In early August the comet will be magnitude 4, but far too close to the Sun to observe, except perhaps in the SOHO LASCO satellite field. After mid-September, 2P/Encke will fade to obscurity very rapidly.

Apart from the possibility of a bright long-period comet being discovered after this *Yearbook* goes to press, Comet 103P/Hartley looks as though it will be the only really easy binocular prospect, in a dark sky, for 2010. The current orbital elements suggest the comet will pass only 0.12 AU from the Earth on 20 October, when it might be as bright as magnitude 5, or possibly 4. Sadly, the Moon will be only three days from full at this time, which will probably eliminate any naked-eye sightings at closest approach. In mid-August 103P should be tenth magnitude as it travels through Pegasus and its brightness should reach seventh magnitude in September in Andromeda. Around the time of its closest approach comet 103P plunges rapidly south through Auriga (Figure 2), crossing Gemini and Monoceros in November. However, as the comet is making such a close flyby of the Earth even a slight change in the date of perihelion will change the comet's path in October quite noticeably. Nevertheless, it does look like a comet worth waiting for, despite the phase of the Moon at the critical time.

Ephemerides for the three brightest cometary prospects of 2010, namely 81P/Wild, 2P/Encke and 103P/Hartley, are given below.

Comet 81P/Wild

Date 2010			2000.0 RA			Dec.		Distance from Earth	Distance from Sun	Elong- ation from Sun	Mag.
0h		h	m	s	°	′	″	AU	AU	°	
Jan	1	12	10	22.6	−0	12	37	1.236	1.681	97.8	10.8
Jan	11	12	31	59.6	−1	59	31	1.135	1.653	102.4	10.6
Jan	21	12	52	52.1	−3	34	46	1.043	1.631	107.1	10.3
Jan	31	13	12	35.4	−4	54	53	0.959	1.614	112.2	10.0
Feb	10	13	30	38.4	−5	56	41	0.885	1.603	117.8	9.8
Feb	20	13	46	20.8	−6	37	26	0.820	1.598	123.9	9.6
Mar	2	13	59	3.5	−6	56	13	0.765	1.600	130.8	9.5
Mar	12	14	8	11.4	−6	53	44	0.722	1.607	138.6	9.4
Mar	22	14	13	21.2	−6	33	04	0.691	1.621	147.2	9.3

Comet 81P/Wild – *cont.*

Date 2010 0h		RA h	m	s	Dec. °	'	''	Distance from Earth AU	Distance from Sun AU	Elong- ation from Sun °	Mag.
Apr	1	14	14	43.1	−6	00	44	0.675	1.640	156.5	9.4
Apr	11	14	13	0.6	−5	25	12	0.676	1.665	165.9	9.5
Apr	21	14	9	30.1	−4	56	16	0.694	1.695	172.2	9.6
May	1	14	5	49.3	−4	42	28	0.732	1.730	167.9	9.9
May	11	14	3	18.7	−4	48	14	0.788	1.769	159.1	10.2
May	21	14	2	54.2	−5	14	31	0.862	1.811	150.0	10.5

Comet 2P/Encke

Date 2010 0h		RA h	m	s	Dec. °	'	''	Distance from Earth AU	Distance from Sun AU	Elong- ation from Sun °	Mag.
Aug	21	11	27	7.5	−1	58	25	1.150	0.504	26.0	7.3
Aug	31	12	41	5.7	−12	05	36	1.114	0.688	37.4	9.3
Sep	10	13	51	58.5	−19	55	13	1.161	0.868	46.5	10.9
Sep	20	14	56	37.0	−25	00	24	1.272	1.037	52.6	12.3
Sep	30	15	52	38.6	−27	48	01	1.427	1.194	55.7	13.4
Oct	10	16	40	0.2	−29	02	48	1.608	1.342	56.3	14.4
Oct	20	17	19	58.9	−29	20	56	1.805	1.481	55.1	15.3
Oct	30	17	54	8.2	−29	05	27	2.009	1.613	52.6	16.1

Comet 103P/Hartley

Date 2010 0h		RA h	m	s	Dec. °	'	''	Distance from Earth AU	Distance from Sun AU	Elong- ation from Sun °	Mag.
Aug	11	22	34	14.7	+27	58	56	0.591	1.482	133.2	10.8
Aug	21	22	40	49.3	+32	30	47	0.491	1.399	134.1	9.9
Aug	31	22	49	20.5	+37	22	13	0.402	1.320	133.7	8.9
Sep	10	23	3	12.5	+42	38	25	0.322	1.247	132.2	8.0
Sep	20	23	30	30.7	+48	27	16	0.251	1.182	130.1	7.0

Comet 103P/Hartley – *cont.*

Date 2010 0h		RA h	m	s	Dec. °	′	″	Distance from Earth AU	Distance from Sun AU	Elong- ation from Sun °	Mag.
Sep	30	0	31	35.8	+54	29	05	0.190	1.129	128.1	5.9
Oct	10	2	44	49.7	+56	06	46	0.142	1.089	126.2	5.0
Oct	20	5	21	33.0	+41	05	38	0.121	1.065	122.1	4.5
Oct	30	6	43	52.2	+17	06	10	0.136	1.059	115.6	4.7
Nov	9	7	19	36.0	+00	13	33	0.175	1.071	113.0	5.3
Nov	19	7	34	42.3	−9	25	32	0.224	1.101	114.6	6.1
Nov	29	7	38	53.2	−14	49	16	0.275	1.146	119.0	6.9
Dec	9	7	36	5.9	−17	31	41	0.328	1.203	124.8	7.7
Dec	19	7	29	9.4	−18	15	03	0.384	1.271	131.3	8.5
Dec	29	7	20	36.9	−17	26	22	0.446	1.346	137.5	9.3

Minor Planets in 2010

MARTIN MOBBERLEY

Over 400,000 minor planets (also known as asteroids) are known. They range in size from hundreds of kilometres in diameter to tens of metres across. Some 200,000 of these now have such good orbits that they have a numbered designation and 15,000 have been named after mythological gods, famous people, scientists, astronomers and institutions.

Most of these objects live between Mars and Jupiter, but some 6,000 have been discovered between the Sun and Mars and 1,000 are classed as potentially hazardous asteroids (PHAs) because of their ability to pass within 8 million kilometres (five million miles) of Earth while having a diameter greater than 100 metres (328 feet). The first four asteroids to be discovered were Ceres (1), Pallas (2), Juno (3) and Vesta (4), which are all easy binocular objects when at their peak because they all have diameters of hundreds of miles. In 2010 Ceres, Pallas and Vesta are well placed (depending on your latitude) and Hebe (6) and Flora (8) are binocular targets too. Ceres will peak at magnitude 7.5 in June/July but will be low down from the UK near the Sagittarius/Ophiuchus border. Pallas will reach magnitude 8.7 in May near the Serpens/Corona Borealis/Boötes borders and Vesta will shine at 6.3 in Leo in February/March. In September/October the asteroid Hebe will peak at magnitude 7.8 in Cetus and Flora will peak during the same months in Aquarius, at 8.3. Ephemerides for these five minor planets, during their best months, are given at the end of this section.

As well as observing bright binocular asteroids, some advanced amateur astronomers with large telescopes and CCDs are often interested in imaging the potentially hazardous asteroids (PHAs), typically hundreds or thousands of metres in diameter, that sail close to the Earth; and also those bigger objects that, while they are not a conceivable hazard, are coming unusually close. In 2010 three PHAs make approaches within 0.05 AU (an AU is the average Earth–Sun distance of 150 million kilometres or 93 million miles), namely 1999 MN, 2001 PT9 and 2002 AC9. In addition three much larger numbered asteroids approach the Earth within 0.2 AU, namely Mithra (4486), Minos (6239) and Epona (3838). The circumstances of these encounters are detailed in the table below, which lists the date and distance at closest approach, the constellation in which closest approach occurs, the speed in arc seconds per minute at closest approach and the peak magnitude. 2002 AC9 only reaches magnitude 18 but may be of interest to the most advanced amateurs with sensitive CCD cameras. Accurate ephemerides can be computed for your location on Earth using the MPC ephemeris service at www.cfa.harvard.edu/iau/MPEph/MPEph.html.

Significant Earth close-approach encounters in 2010

Minor planet	2010 Date	Distance (AU)	Constellation	Speed	Magnitude
2001 PT9	Mar 2.10	0.03532	Cen./Crux	32″/min	15.4
Mithra	Mar 12.57	0.1890	Ophiuchus	10″/min	14.9
1999 MN	June 4.48	0.03342	Sagittarius	35″/min	15.7
Minos	Aug 10.88	0.09849	Cetus	9″/min	15.8
2002 AC9	Aug 13.14	0.04716	Centaurus	24″/min	18.0
Epona	Nov 7.31	0.1973	Cygnus	12″/min	14.7

Ephemerides for the best-placed/brightest minor planets at opposition in 2010

1 Ceres

Date 2010		2000.0 RA			Dec.			Geocentric Distance	Heliocentric Distance	Elongation	Visual Magnitude
		h	m	s	°	′	″	AU	AU	°	
May	1	18	19	23.6	−22	39	12	2.091	2.804	125.9	8.1
May	11	18	17	32.1	−23	10	02	1.996	2.812	136.0	7.9
May	21	18	12	58.6	−23	44	34	1.918	2.819	146.7	7.7
May	31	18	06	01.2	−24	21	06	1.862	2.827	157.9	7.5
Jun	10	17	57	14.2	−24	57	14	1.831	2.835	169.4	7.3
Jun	20	17	47	30.0	−25	30	31	1.827	2.842	177.5	7.1
Jun	30	17	37	53.0	−25	59	14	1.851	2.850	166.8	7.4
Jul	10	17	29	23.4	−26	22	58	1.901	2.857	155.4	7.6
Jul	20	17	22	51.4	−26	42	26	1.975	2.864	144.5	7.9
Jul	30	17	18	48.4	−26	58	59	2.071	2.871	134.1	8.1

2 Pallas

Date 2010			2000.0 RA				Dec.		Geo-centric Distance	Helio-centric Distance	Elong-ation	Visual Magni-tude
		h	m	s	°	′	″		AU	AU	°	
Mar	11	15	51	31.9	+10	47	07		2.113	2.667	113.4	8.9
Mar	21	15	53	58.7	+13	33	15		2.044	2.692	120.6	8.8
Mar	31	15	53	45.1	+16	23	14		1.992	2.716	127.3	8.7
Apr	10	15	50	53.2	+19	08	13		1.959	2.740	132.8	8.7
Apr	20	15	45	37.3	+21	37	58		1.947	2.764	136.6	8.6
Apr	30	15	38	30.4	+23	42	25		1.957	2.788	138.0	8.7
May	10	15	30	18.9	+25	14	09		1.989	2.812	136.9	8.7
May	20	15	21	57.0	+26	08	53		2.041	2.835	133.6	8.8
May	30	15	14	20.0	+26	26	38		2.112	2.858	128.8	8.9
Jun	9	15	08	10.3	+26	10	56		2.199	2.881	123.1	9.1

4 Vesta

Date 2010			2000.0 RA				Dec.		Geo-centric Distance	Helio-centric Distance	Elong-ation	Visual Magni-tude
		h	m	s	°	′	″		AU	AU	°	
Jan	11	10	41	43.6	+14	54	09		1.644	2.432	133.9	6.9
Jan	21	10	39	17.6	+15	56	43		1.551	2.422	144.7	6.7
Jan	31	10	33	57.7	+17	13	57		1.480	2.413	156.0	6.5
Feb	10	10	26	09.7	+18	38	57		1.432	2.403	166.7	6.2
Feb	20	10	16	45.9	+20	02	09		1.412	2.394	171.0	6.1
Mar	2	10	07	03.3	+21	13	27		1.420	2.384	162.6	6.3
Mar	12	9	58	22.0	+22	05	39		1.453	2.374	151.5	6.5
Mar	22	9	51	50.1	+22	35	05		1.510	2.364	140.4	6.6
Apr	1	9	48	11.0	+22	41	46		1.585	2.354	130.0	6.8
Apr	11	9	47	38.2	+22	28	04		1.674	2.345	120.3	7.0

6 Hebe

Date 2010		2000.0 RA			Dec.			Geo-centric Distance AU	Helio-centric Distance AU	Elong-ation °	Visual Magni-tude
		h	m	s	°	′	″				
Aug	10	0	29	41.4	−07	52	35	1.166	1.997	132.6	8.5
Aug	20	0	33	00.8	−09	59	08	1.090	1.984	141.4	8.2
Aug	30	0	33	19.8	−12	28	53	1.032	1.973	150.2	8.0
Sep	9	0	30	42.0	−15	10	50	0.993	1.963	157.7	7.8
Sep	19	0	25	41.0	−17	48	04	0.977	1.954	161.0	7.7
Sep	29	0	19	22.7	−20	02	11	0.983	1.947	157.7	7.7
Oct	9	0	13	08.6	−21	38	29	1.010	1.941	150.3	7.9
Oct	19	0	08	23.2	−22	29	05	1.056	1.937	141.5	8.1
Oct	29	0	06	07.5	−22	34	25	1.118	1.935	132.7	8.3
Nov	8	0	06	50.0	−21	59	37	1.194	1.934	124.3	8.6

8 Flora

Date 2010		2000.0 RA			Dec.			Geo-centric Distance AU	Helio-centric Distance AU	Elong-ation °	Visual Magni-tude
		h	m	s	°	′	″				
Jul	31	23	49	06.8	−08	29	59	1.172	2.007	133.1	9.2
Aug	10	23	50	01.9	−09	26	31	1.089	1.991	142.5	9.0
Aug	20	23	47	42.5	−10	43	19	1.021	1.975	152.6	8.7
Aug	30	23	42	21.6	−12	13	36	0.973	1.960	162.6	8.4
Sep	9	23	34	39.8	−13	46	34	0.947	1.946	169.5	8.2
Sep	19	23	25	54.6	−15	08	22	0.943	1.933	165.8	8.2
Sep	29	23	17	43.3	−16	06	51	0.961	1.920	156.0	8.4
Oct	9	23	11	33.1	−16	34	53	1.000	1.909	145.5	8.7
Oct	19	23	08	28.0	−16	30	30	1.055	1.898	135.4	8.9
Oct	29	23	08	53.3	−15	56	21	1.124	1.889	126.1	9.1

Meteors in 2010

Meteors ('shooting stars') may be seen on any clear moonless night, but on certain nights of the year their number increases noticeably. This occurs when the Earth chances to intersect a concentration of meteoric dust moving in an orbit around the Sun. If the dust is well spread out in space, the resulting shower of meteors may last for several days. The word 'shower' must not be misinterpreted – only on very rare occasions have the meteors been so numerous as to resemble snowflakes falling.

If the meteor tracks are marked on a star map and traced backwards, a number of them will be found to intersect in a point (or a small area of the sky) which marks the radiant of the shower. This gives the direction from which the meteors have come.

The following table gives some of the more easily observed showers with their radiants; interference by moonlight is shown by the letter M.

Limiting Dates	Shower	Maximum	Radiant RA h	m	Dec. °	
1–6 Jan	Quadrantids	3 Jan, 18h	15	28	+50	M
19–25 April	Lyrids	21–22 April	18	08	+32	
24 Apr–20 May	Eta Aquarids	4–5 May	22	20	−01	M
17–26 June	Ophiuchids	19 June	17	20	−20	
July–August	Capricornids	8, 15, 26 July	20	44	−15	
			21	00	−15	
15 July–20 Aug	Delta Aquarids	29 July, 6 Aug	22	36	−17	M
			23	04	+02	M
15 July–20 Aug	Piscis Australids	31 July	22	40	−30	M
15 July–20 Aug	Alpha Capricornids	2–3 Aug	20	36	−10	
July–August	Iota Aquarids	6–7 Aug	22	10	−15	
			22	04	−06	
23 July–20 Aug	Perseids	12 Aug, 21h	3	04	+58	
16–31 Oct	Orionids	20–22 Oct	6	24	+15	M
20 Oct–30 Nov	Taurids	2–7 Nov	3	44	+22	
			3	44	+14	

Limiting Dates	Shower	Maximum	Radiant			
			RA		Dec.	
			h	m	°	
15–20 Nov	Leonids	18 Nov, 00h	10	08	+22	M
Nov–Jan	Puppid-Velids	early Dec	9	00	−48	
7–16 Dec	Geminids	14 Dec, 06h	7	32	+33	
17–25 Dec	Ursids	22 Dec	14	28	+78	M

Some Events in 2011

ECLIPSES

There will be six eclipses, four of the Sun and two of the Moon.

4 January:	Partial eclipse of the Sun – Europe, northern Africa, Middle East, southern Asia
1 June:	Partial eclipse of the Sun – Alaska, northern Canada, Greenland, Iceland, north-eastern Asia
15 June:	Total eclipse of the Moon – Europe, Africa, southern Asia, Australasia
1 July:	Partial eclipse of the Sun – Antarctica
25 November:	Partial eclipse of the Sun – Antarctica, South Island of New Zealand, Tasmania, South Africa
10 December:	Total eclipse of the Moon – North America, Europe, Africa, Asia, Australasia

THE PLANETS

Mercury may be seen more easily from northern latitudes in the evenings about the time of greatest eastern elongation (23 March) and in the mornings about the time of greatest western elongation (3 September). In the Southern Hemisphere the corresponding most favourable dates are 20 July (evenings) and 7 May (mornings).

Venus is at greatest western elongation on 8 January. It is visible in the mornings from the beginning of the year until early April for Northern Hemisphere observers, and until early July for those in the Southern Hemisphere. The planet passes through superior conjunction on 16 August, and then becomes visible in the evenings towards the end of the year.

Mars does not come to opposition in 2011. The planet is in conjunction with the Sun on 4 February, and becomes visible in the morning sky later in the year.

Jupiter is at opposition on 29 October in Aries.

Saturn is at opposition on 3 April in Virgo.

Uranus is at opposition on 26 September in Pisces.

Neptune is at opposition on 22 August in Aquarius.

Pluto is at opposition on 28 June in Sagittarius.

Part II

Article Section

An Introduction to Visual Planetary Astronomy

PAUL G. ABEL

In the last few years we have witnessed something of a revolution in amateur astronomy. No one can deny the benefits that both the CCD and the webcam have made to the lives of amateur astronomers. The use of these devices has proliferated down into all aspects of modern astronomy – so much so, in fact, that the newcomer could be forgiven for thinking that the only way to make a start in amateur astronomy today is to buy a powerful computer, a CCD, a webcam and, finally, a telescope. This view is no doubt reinforced by the exuberant supply of stunning images on the pages of our popular astronomy magazines.

Like all revolutions, there is a price to pay. That price has been the slow decline in the number of visual observations and visual observers who have traditionally played, and must continue to play, an important rôle in amateur astronomy. I should make it clear at this point that I'm in no way anti-technology; nor is this chapter in any way intended to play down the significant contributions of the CCD astronomer and the sterling work of astronomers such as Damien Peach, John Fletcher and many more. No, rather this article is here simply to re-address the balance, there now being a wealth of literature on CCD astronomy. In particular I hope to show that one can make very accurate planetary drawings quite inexpensively, and that by sketching these worlds we can achieve the degree of understanding that accompanies seeing. I think it is fair to say that many images made at the telescope these days simply sit on the hard drive using up space. Armed with a little visual understanding, one can make better images and better drawings of real use, thus elevating the amateur astronomer closer to his professional counterpart.

This article is not just for amateurs who want to contribute their observations to the various scientific organizations. It is also for the amateur who has a telescope and simply wishes to go out and look; by learning some of the tips of visual astronomy you will learn to see things that may have escaped your attention before!

I have broken up this chapter into six sections. In the next section we shall look at the 'hows' and 'whys' of visual astronomy. We see why we still make visual observations and learn about the general tools of the trade. The four sections after that are devoted to the planets that can be observed by anyone with a telescope, namely Venus, Mars, Jupiter and Saturn. I have omitted Mercury, as it is never far from the Sun and, in any case, most people probably won't catch it. I have also omitted Uranus and Neptune: these worlds are rather remote and although they're visible in binoculars, they require a large aperture if any detailed observations are to be made. Finally, I have left out my favourite astronomical body, the Moon. I have done this simply because sketching the Moon would be a chapter in itself! I would also be hindered by the fact that I'm still developing techniques for sketching all the different shadows, craters and mountains. Truly the lunar surface is a test of any amateur's skill!

So, let us make a start. I think a good place to begin is by asking why should we want to do visual astronomy, given recent advances, and how can we get the best results out of looking through the eyepiece?

THE HOW AND WHY OF VISUAL ASTRONOMY

One of the problems visual astronomy is commonly saddled with (as is sketching at the telescope) is the perception that it is not as accurate as photography or imaging. Not only is this statement false, but it demonstrates a curious lack of faith in something that has served mankind well for generations: the human visual system. Not only is the human visual system capable of producing an accurate representation of the physical world (without its accuracy we should not be able to cross roads, judge heights, assess risk and so on), but further, it can be trained to be a precise piece of astronomical equipment.

To demonstrate this, I will relate an experiment I carried out back in 2008. That year, Saturn was well placed in the evening sky and I decided to invite some friends (most of whom had never looked through a telescope, or realized that my all-consuming hobby had gone so far!) over to my observatory in Leicester, which houses my 8-inch reflector. I wondered how these novices might fare if I gave each of them an observing form and asked them to sketch Saturn as seen through the telescope. Each of them watched, bemused, as I issued them with clipboards containing a Saturn outline and strict instructions to give me their sketches after their observation and not to discuss with anyone what they had seen at the eyepiece. The first volunteer was deposited at the eyepiece of the telescope and, after being warned he would have only twelve minutes to make a sketch, left to record his view through the eyepiece. To get an objective view of what was seen I also got a few CCD astronomers to image Saturn so that I could compare all our sketches.

The results were startling, to say the least. Not only had people recorded the Southern Equatorial Belt (a vast band crossing the globe), the dusky polar regions and the Cassini Division in all of the correct positions: many of them had correctly recorded details we would consider to be quite difficult to pick out and, moreover, some of those details were vague in the images. Details like the doubling of the Southern Equatorial Belt (it appeared as a single feature on the image), the dark C-Ring, the shadow of Globe on Rings and Rings on Globe along with a bright equatorial zone all turned up on the sketches. What was even more interesting was the fact that most people thought the CCD image was not a fair representation of what they had seen at the eyepiece, even though it was taken at about the same time. It is interesting to note that one member of our group was an artist. I have to confess I had thought that being an artist he would exaggerate everything beyond all proportion and produce an image that the Cassini probe would be proud to own; not so! His sketch contained the least detail, thus demonstrating that artistic ability need not cause people to exaggerate their observations.

Of course, I don't claim this experiment to be the last word in measuring the accuracy of visual observers, but what it does show is that even people who have never looked through a telescope before, and have no knowledge of the subject, can produce a decent drawing representing what they have seen. It is clear, then, that given time, we can train ourselves to really see what is at the eyepiece; the amateur who pursues visual astronomy will learn how to see all over again and will soon discover that looking is not the same as seeing.

Learning to see what is in the eyepiece and recording it as accurately as possible is the aim of visual astronomy, for pursuing these goals we become better placed to judge accurately what our eyes are telling us and to interpret it in a scientific way. Any unusual phenomenon that might manifest itself in the atmosphere of one of the planets becomes startlingly obvious, and after a few weeks of examining a planet visually, you will find that the surfaces of these worlds can become as familiar as the London Tube or Selsey Bill.

I want to dispel another myth about planetary sketching, this being that you somehow need to have artistic talent to draw the planets. The reason this is untrue is that our aims are very different from those of an artist. We wish to record as objectively and as accurately as possible what is and is not there to the best of our abilities. It matters very much where we place a feature and that we correctly record its appearance. We cannot neglect something because it doesn't look or feel right, and we are not free to represent what we have seen in any way we wish. This is the difference between artistic sketching and draftsmanship.

The next question that needs to be answered is 'what do we need?' To keep things simple, I shall be concerned only with producing black-and-white sketches here. With this in mind, I recommend the following items:

- a collection of pencils, including 2B, 3B and 4B pencils;
- white correcting fluid (great for bright clouds and features);
- a clipboard;
- a red light;
- artist's smudge stick (or cotton-wool buds)
- most important: a sturdy hardback notebook.

Recording Your Observations. I don't think the importance of recording observations can be stressed enough. The act of writing up what you have seen – mentioning, for example, any new techniques you've tried to record a crater correctly, or which eyepiece is of no use for Venus – crystallizes the time at the telescope in your mind. You will find that your observing logs become fascinating things in themselves – a complete record of your first sights at the telescope to your latest more sophisticated observations. I don't mind admitting that I like looking back at my early sketches and reading about those early triumphs (and disasters!), and I never tire of reading Patrick Moore's observing books.

I shall discuss the specifics of what to record and how to sketch each of the planets shortly; however, in general, there are two steps in recording your observations:

1. *Rough sketch made at the telescope.* This contains your rough drawings and any comments about the appearances of the object and its features.
2. *Fair copy.* This is the neat and tidy version of your notes made in your log book. I have a log book for each planet, and I have included a page from my Mars book in Figure 1 to give some idea of the general layout.

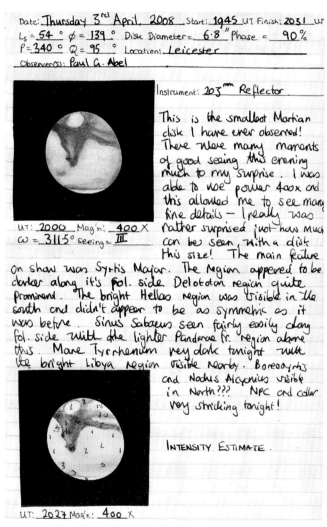

Figure 1. A page from the author's Mars log book.

Remember to include in both the rough and final version the date, the start and finish time (in UT) of your observations, along with the details of telescope used (i.e. size and type), magnification and seeing conditions. Seeing is normally measured on the Antoniadi scale ranging from I (perfect seeing, seldom achieved) to V, which is the worst type of seeing – the object appearing to 'boil' in the eyepiece.

I don't propose to get into a discussion here about telescopes. Suffice it to say that the best telescope to use is the one that suits you and your circumstances. That's equipment covered. Let's now turn our attention to our nearest neighbour.

VENUS

Venus has something of a reputation amongst astronomers for showing very little detail. While it is certainly true that details in the Venusian clouds are rather low contrast, there is no truth in the statement that the Venusian disk is bland. Interesting low-contrast regions stride the disk, often surrounding brighter regions; with a little practice, the visual observer can soon train his eye to see the elusive markings. It is also interesting to watch the phase increase or decrease as Venus moves closer/further away from us.

Observing Venus. Whenever we look at Venus through a telescope, all we are ever observing is the top layer of a rather extensive cloud deck which rotates about the planet every four-and-a-half days or so. We have mapped the surface of Venus in great detail using radar and spacecraft, since the poisonous clouds never clear away. These clouds are not just a bland veil shrouding the planet. They manifest all manner of interesting subtle changes in intensity and structure. Although features come and go, there are some standard features associated with the disk and these are demonstrated in Figure 2. They are as follows:

- *Bright limb.* This is the bright outer region of the disk and is normally the brightest feature on the disk. It is usually complete; however, on occasions it can be broken up or irregular. The bright limb tends to fade into the dusky markings; however, if Venus is observed in a dark sky, the bright limb may be overwhelming and wash out all but the most striking of markings, so it is best to observe the disk in a bright sky.

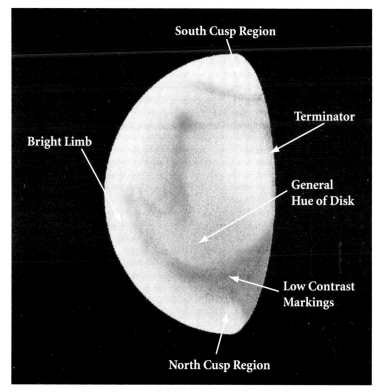

Figure 2. Some of Venus' long-standing features. Drawing by the author.

- *Terminator.* The boundary between the night and daytime. Usually the terminator is a nice regular feature. However, it can appear to be quite irregular, as is the case in Figure 3. Very often there are dark regions on the terminator which can fade into the general disk hue.
- *Cusp regions.* Very often these manifest themselves as bright regions at the poles. They are normally of equivalent brightness; sometimes, however, one cusp can appear brighter than the other. The cusps are normally bordered by the cusp collars, distinct polar bands just below the polar cusps.
- *Cusp extensions.* As Venus comes closer to us, so its phase decreases. As it does so, the light in the atmosphere is scattered in such a way that the bright limb appears to extend into the night region, giving rise to what are known as cusp extensions, as shown in Figure 4.

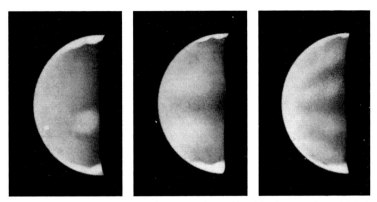

Figure 3. The irregular terminator of Venus as observed by Richard Baum using a 4.5-inch refractor.

Figure 4. Cusp extensions on Venus observed by Richard Baum with a 4.5-inch refractor.

It is normally best to observe Venus in a light or twilight sky, as this will reduce the glare. If you wish, you could try a neutral density filter; this will certainly reduce the glare.

When it comes to magnification, I have always found that a power of 200x is sufficient. In my experience, Venus does not tolerate high power well. When making a visual observation of Venus, we should have the following three main objectives in mind. First we need to get a disk drawing. Secondly, make an intensity estimate for each of the features you have observed. Finally, try to make a phase estimate. This may sound a little pointless; the phase is, after all, printed in the *Handbook of the British Astronomical Association* and a dozen other places. However, there a difference between the predicted phase and the actual phase, because of a phenomenon known as the Schröter effect (the scattering of light in the thick Venusian atmosphere). To make a phase estimate, take your disk drawing (see below) and measure the x distance as shown in Figure 5. The y distance should always be 50 mm, and a simple formula gives the phase:

$$\text{Phase} = x / 50$$

Thus, if you measure x to be 30 mm, then the phase is 30/50 = 60 per cent.

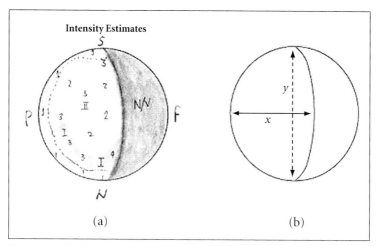

Figure 5. On the left an intensity estimate of Venus; on the right the distances x and y.

Disk Drawing. For this you will need pencils, a Venus blank – which should be a circle of 50 mm diameter, perhaps printed on a black background – and a clipboard to lean on. Once you have Venus in the eyepiece, take a little time to see what's there. You should ask yourself where the terminator is, what the appearance of the bright limb is and what features can be seen, and so on. You'll find that by asking yourself such questions, your eyes will seek out the answers.

The next thing to do is to make a sketch. There are many different ways of completing a sketch and I cannot hope to give them all here, so I will tell you what I do. The first thing to do is draw in the terminator. Draw this as accurately as you can, as you will need it to make the phase estimate. Next, put down all prominent features you can see. These may be the cusp collars and any markings along the terminator; you will see that these features stand out from a general dull shading present everywhere else. A 2B pencil is good for these markings. This shading will fade into the bright limb band. You might want to smudge with your finger all the features that appear vague or without a definite boundary. Finally, note the time of the sketch. At this point I always check the features in the sketch with what I can see in the eyepiece. I find it easier to do it one feature at a time, so I check that the markings are how they appear in the eyepiece, the location of any bright regions and so on.

Intensity Estimates. In essence, an intensity estimate is a disk drawing with numbers marking the intensity of each feature or region. With Venus, we use a scale of 0 to 5: 0 indicates extremely bright regions, 1 indicates bright areas, 2 indicates the general hue of the disk, 3 says that the features are just on the edge of visibility, a 4 indicates the shadings are well seen, while finally a 5 indicates unusually dark shadings. I have included an intensity estimate in Figure 5. To make an intensity estimate, first make another disk drawing, then assign an intensity number (0–5) for each of your features. In Figure 5 we see that the limb band was very bright, taking a value of 0, while some of the shadings in the south were just on the edge of visibility, hence a 3, and so on.

Ashen Light. I cannot finish this section without mentioning something about the Ashen Light. I must say I have never observed this feature myself, but I know many good visual observers who have. The

Figure 6. A drawing of the crescent Venus by David Gray showing the Ashen Light.

Ashen Light can be observed only when Venus is in its crescent phase and in a dark sky, and it takes the form of a faint glow, illuminating the night side of the planet. David Gray's sketch in Figure 6 shows how the light gives structure to the nightside of Venus. The cause of the Ashen Light remains tantalizingly vague, at the time of writing; the favoured theory is that it may be caused by a thinning of the clouds on the nightside which, as a result, is allowing us to see the warm surface below.

MARS

Mars is rather unaccommodating to observers with small telescopes. Oppositions occur once every two years, during which time Mars may be only a disappointing 4 arc seconds across. It is only some few months before and after opposition that Mars really becomes favourable. Moreover, because of the eccentricity of Mars's orbit, not all oppositions are equally favourable; oppositions occurring when Mars is at aphelion (i.e. furthest from us, as is the case for 2010) mean that Mars is unlikely to exceed 14″ in size. However, when an opposition occurs at perihelion, the disk diameter may be as large as 25″. There is

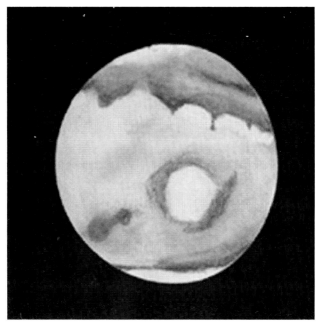

Figure 7. Mars as seen through Patrick Moore's 12.5-inch reflector. Drawing made by the author with Roger Prout.

some compensation for aphelion oppositions, however: during this time Mars is very high in the sky for Northern Hemisphere observers and so is much less affected by poor seeing. At perihelion Mars will be very low in the sky for Northern Hemisphere observers (as will be the case in July 2018).

Mars is a wonderful world to behold in the eyepiece, and of all the worlds in the Solar System, Mars is surely the one with which we have the closest affinity. Here is a world remarkably similar to ours, with an atmosphere (albeit a rather thin one), water ice, seasons and volcanoes. Even with a power of 130x on a 4.5-inch reflector around opposition, I have been able to see the polar ice caps and dark surface markings. I had my best views of Mars so far in 2008 through Patrick Moore's 12½-inch reflector: the amount of detail on show was truly remarkable (as can be seen in Figure 7). Whatever the size of telescope, during opposition week you will see something!

Observing Mars. A visual observation of Mars should consist of a disk drawing and an intensity estimate. There is no need for a phase estimate, since the Martian atmosphere is very thin and Schröter effect is negligible here. There is some essential preparatory work to be done before going to the telescope, however; we shall need to prepare a correctly orientated observing blank first.

The blank needs to be the correct size, 50 mm. We now need to get the phase, the position of the north pole (the angle P) and the position of greatest defect of illumination (Q). I have illustrated these angles in Figure 8. For the night of 1 February 2008 the *Handbook of the British Astronomical Association* gives P = 331° and Q = 92° with a phase of 95 per cent. This gives us a blank, as shown in the figure.

As with Venus, there are some standard things to look out for when making an observation of Mars. These include the following:

- *Observation of the polar ice caps.* Both the northern and southern polar ice caps shrink during their respective summers and increase again during the winters. Thus, spring times will see the ice evaporating and hence mist and fogs in the polar regions. Such mists and fogs will look like bright white regions obscuring the darker features underneath. The caps do not always expand and contract at the same

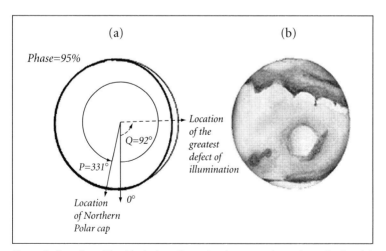

Figure 8. Preparing a Mars blank. As well as using the correct phase, the angles P and Q need to be taken into account.

rate, so it is interesting to measure how quickly they evaporate and re-form year after year.

- *Dust storms.* Mars is a very dusty sort of world, and although the atmosphere is tenuous, there is enough of it to create winds, and where there are winds there are dust storms. A dust storm can start up at any time and usually manifests itself as yellowish cloud obscuring some particular region. A dust storm can rapidly spread over the whole of the surface, blowing a great deal of dust into the atmosphere. This has the undesirable effect of obscuring all of the disk markings. Figure 9 shows a dust storm in the great Valles Marineris region.

- *White clouds.* White clouds may form over the volcanoes in the Tharsis region and the high plateaux. They are usually very bright and can often spill over the terminator into the nightside. They can be quite stunning.

Disk Drawing and Intensity Estimates. Equipped with a correctly orientated blank, we are ready to attempt a disk drawing. We have to be a little careful here. Mars will have rotated more than 3.5°, so we have only fifteen minutes to make a drawing. After this time Mars's rotation will have moved the features significantly from where we first placed them on paper.

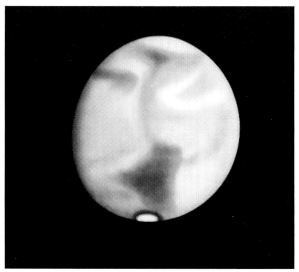

Figure 9. A dust storm is forming over the Valles Marineris region of Mars. Drawing by David Gray using a 415 mm Dall-Kirkham telescope.

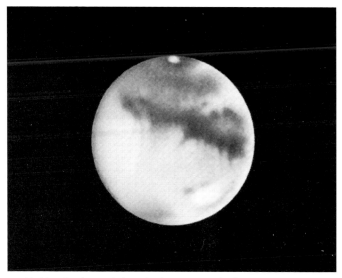

Figure 10. Mars drawing made by Richard Baum. Note the fine details which can be seen just using a 4.5-inch refractor when the disk is sufficiently large.

The first features I always draw in are the polar ice caps. Next, draw in all the prominent regions along with the dark markings and bright regions. If there are any details on the following or preceding limbs, record these first, as they will be most affected by the rotation. I always draw in the most obvious feature first and put the other features on in relation to this to get the correct positions. Once you have done this, note down the time (in UT). Finally, you can add the finishing touches: smudge any of the surface features which have ill-defined edges, note the boundaries of any dark or bright regions or white clouds. You may want to write some notes near the drawing to make sure you redraw them correctly when you copy them up neatly later on. During opposition week, an average-sized telescope on good nights might well reveal a wealth of fine detail, as was the case for Richard Baum, whose drawing is given in Figure 10.

To make an intensity estimate, you need to make a disk drawing in the same manner as above, but this time we assign each of the features a number between 0 and 10, 0 being a bright white cloud and 10 the dark sky. We use this scale for Mars, Jupiter and Saturn.

You will find Mars a fascinating world, and once you have drawn the whole planet you can make yourself a map to show how the entire surface appears according to your observations. It is always great fun to compare your map with an official one to see how well you did.

JUPITER

Jupiter is surely every amateur's best friend! The disk is so large and bright (47 arc seconds at opposition) that it positively defies the eye not to see anything on it. My first view of Jupiter was through a 4.2-inch reflector and with this I was able to see about four belts, the bright zones, the dusky poles and, of course, the four bright Jovian satellites. I remember being quite mesmerized by their eternal encounters as they danced around the globe.

Jupiter is a very different kind of world from those of Mars and Venus. We class it as a gas giant and it is predominantly made of hydrogen and helium. When we observe Jupiter, we are viewing the top layer of an extremely thick and tumultuous atmosphere, where winds rage at a staggering 300 km/s and storms the size of the Earth have existed for hundreds of years, there being no solid surface to dissipate them. The lack of a solid surface means that Jupiter's equator rotates more quickly than the rest of it and, as a result, it bulges out considerably and appears appreciably flattened through a telescope.

Jupiter comes to opposition every thirteen months and spends about one year in each constellation. For UK observers, the past few years have been disappointing, with Jupiter languishing very low in the south. Since last year, however, things have been improving and in 2010, Jupiter comes to opposition on 21 September in the constellation of Pisces. When Jupiter is high in the sky, it is possible to follow one rotation during an entire night. This is made possible by the fact that Jupiter rotates once in about ten hours or so.

Observing Jupiter. It is worthwhile getting to know the names and locations of the various Jovian features. Figure 11 shows some of the standard belts and zones that are well within the grasp of a 6-inch reflector. Because of the different rotational speeds of the planet, astronomers employ the use of two longitude systems, creatively called System I and System II. System I consists of all the features between the NEBs and SEBn (including the equatorial zone) and has a rotational

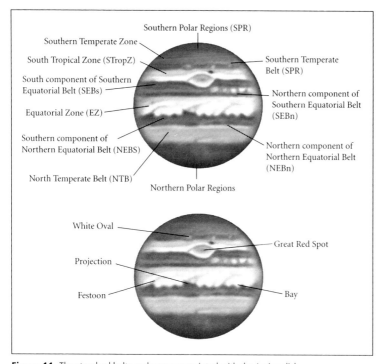

Figure 11. The standard belts and zones associated with the Jovian disk.

period of 9h 50m. System II consists of everything else and takes slightly longer to rotate, 9h 55m. There is a further system, System III: this is the rotational period of Jupiter's magnetic core.

An inspection of Jupiter at about 180x or so will reveal the edges of the belts to be mottled because of the existence of white ovals, dark projections and festoons. These features are well worth recording, but it can be a little difficult to recognize these many different weather systems, and so Figure 11 demonstrates some of the most regularly reported features of the Jovian clouds. Really good views of Jupiter can be had with an 8-inch reflector. Even in poor conditions I have always been able to observe plenty of features on the disk. This is evident in my drawing in Figure 12, made in 2008 when Jupiter was very low in the south.

A visual observation of Jupiter should have the following aims:

Figure 12. A view of Jupiter through an 8-inch reflector in less than ideal conditions! Drawing by the author.

Longitude estimates. The various features like white spots and so on are not fixed in longitude; rather they drift around in longitude and the speed of this drift depends largely on the circulation currents feeding these spots. Determining the longitude of a feature is a simple procedure; all we do is wait until the feature lies on the central meridian. We then record the time (in UT) that the feature appeared on the CM. We can then convert this time into a longitude, and by repeating this throughout the apparition, we may obtain a series of longitude estimates which will show how quickly the feature is drifting in longitude.

Disk Drawing. When making a disk drawing of Jupiter, we have even less time to complete it than we did for Mars. In fact, in order for the sketch to remain true it must be completed within ten minutes. At first this seems a little daunting, especially when you are faced with a view of Jupiter at 200x or so; however, with practice, a good sketch can be accomplished within this time.

When going out to the telescope, make sure you have a blank that correctly represents Jupiter's oblateness – a standard blank has a width of 63 mm and a height of 59 mm. As usual, spend about five to ten minutes taking in the sights. You should see how many belts and zones are visible, and then determine what activity you can see within the various belts and zones.

Once you are reasonably confident of what is there, it is time to put pencil to paper. I normally draw in all the belts I can see. Next, put in all the details close to the following and preceding edges of the disk. These features may be irregularities in the belts, white spots and so on. Next, put down the locations of all the other spots, projections and any other atmospheric disturbances. This should take about six or seven minutes, and once you have finished, note the time of your sketch. Finally, use the last few minutes to put in the fine details and smudge the edges of the belts and zones. Now's a good time to write any notes about the location and appearance of the various features you have observed. Finally, check that you have recorded everything as accurately as you can.

Although it may no longer have any scientific value, it is nonetheless rather magical to watch the Jovian satellites passing in front and behind

Figure 13. Satellite transiting the Jovian disk. Drawing by David Gray with a 415 mm Dall-Kirkham.

the giant planet. It is always worth sketching the transit of a satellite, and with larger telescopes, you may be able to determine the colour of the satellites as they pass across the disk. David Gray's sketch in Figure 13 shows a satellite transiting Jupiter, followed by its shadow.

Intensity Estimates. When making an intensity estimate for Jupiter, we use the same scale as we did for Mars, 0–10. The first step is to make a disk drawing as described above, and then at the end assign the various features the correct intensity values. Come the end of the apparition, it is worth tabulating your results to see if any of the features have fluctuated in intensity over the course of the season. Whatever the size of your telescope, Jupiter will keep you busy!

SATURN

If you really want to impress your friends and give them some idea of why astronomy is so absorbing, all you have to do is show them Saturn through a telescope, for it is surely one of the most visually stunning and intriguing sights a human being can see. I well remember colleagues asking the inevitable question, 'Why the hell on Earth do you risk hyperthermia looking at stars?' The only thing to do was show them Saturn at 400x through my 8-inch reflector. After a few hours I managed to prise them away and get them back indoors . . .

Saturn, like Jupiter, is classed as a gas giant. That superb ring system is not solid, of course. It is composed of millions of particles ranging in size from small pebbles to enormous boulders. Although Jupiter and Saturn are of a similar make, the striking weather patterns for which Jupiter is justly famous are much less evident on Saturn. Although Saturn has a very dynamic atmosphere, there is a layer of fog and mist over it. As a result, Saturn presents a much calmer face to us, although great weather systems can and do break out. Every fifty-seven years or so a great white spot erupts in the equatorial zone. Smaller spots occur quite frequently. In 2008, I was busy monitoring two white spots in the south tropical zone.

Saturn takes almost thirty years to complete one circuit around the Sun and oppositions occur once a year, each opposition being approximately thirteen days later. As a result, Saturn is a slow mover in the sky and may reside in a particular constellation for a number of years.

Figure 14. (a) The rings of Saturn at a very shallow angle but still visible. Drawing by the author; (b) Drawing by Richard Baum where the rings are invisible.

Saturn has an axial tilt and so, as it journeys around the Sun, the rings are on display at different angles: some years they are wide open, and at other times they are edge on (as was the case last year, 2009). When the rings are viewed edgeways on they can seem to disappear altogether, as was the case when Richard Baum made his sketch of the planet in Figure 14. In the early part of 2008 the rings were at a very shallow angle, giving Saturn a most unusual appearance, as is evident in my drawing (also in Figure 14).

Observing Saturn. Saturn displays similar, though less distinct, weather systems to those of Jupiter. Like Jupiter, there are established surface

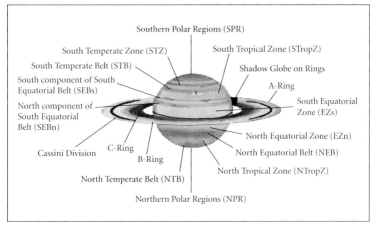

Figure 15. Some of the regular features seen on Saturn.

features which I have given in Figure 15. It has to be said that the various belts and zones of Saturn are much more muted, and determining where a belt ends and zones begin requires some practice!

Saturn also has three systems of longitude. System I, which applies to features in the equatorial zone, NEB and SEB, has a rotational period of 10h 14m. System II, which has a rotational period of about 10h 38m, is for the rest of the globe, excluding the NPR and SPR. For the polar regions, System I is again used. (Finally, we have System III, which is Saturn's radio rotation rate: this was measured by Voyager as 10h 39m, but by Cassini as 10h 45m.) We can calculate the longitude of features for Saturn in exactly the same way as we do for Mars and Jupiter. Tables of longitude can be found in the *Handbook of the British Astronomical Association.*

A full visual observation should comprise the following parts:

Disk Drawing. Saturn is perhaps one of the hardest planets to convey on paper. The first difficulty is representing Saturn correctly. The oblateness of the disk, the size of the rings and their correct angle as viewed from Earth at the time must all be correctly represented. Fortunately, the *Sky at Night* magazine produces an excellent CD containing the correct observing blanks.

Since Saturn rotates rather quickly (by terrestrial standards), we must aim to complete a sketch of the planet in about ten minutes. The first thing to do is to sketch in the darkest features. This will almost certainly be the SEB, if Saturn's southern face is presented to us (or the NEB if its northern face is on show). Start by shading in the most prominent belt. Very often the SEB will appear as two belts, with a darker and thicker SEB and a thinner SEB. In apertures of 4 inches or larger the STB should be visible and the poles should appear dark and dusky. Similarly in the north, the NTB should be visible as a thin belt and the NPR should be shaded similarly to the SPR. The EZ tends to be very bright, and occasionally a thin incomplete EB can be seen. I tend to use a 3B pencil for the main belts.

Once you have completed the disk drawing, move on to the rings. The brightest ring is the B-ring and larger apertures (say 6 inches or greater) will often show variations including a darker B2-ring – some gentle shading with an HB or B pencil will help convey the subtle changes. Next in brightness is the A-ring. As a rule, I normally gently shade in the A-ring with a B pencil and paint a small quantity of water over it. The final ring to be represented is the C-ring, which may be seen only in apertures of 4 inches or more when the rings are quite wide open. Unlike the other two, this ring is dark and semi-translucent. For the C-ring, use a heavy pencil to shade it in. Smudge the pencil to achieve the transparency effect associated with the C-ring.

The final task is to record the Cassini Division, if it can be seen. Encke's division should be visible if the rings are open sufficiently; I have seen it quite easily with my 8-inch reflector. The very last thing you need to mark on is the Shadow of the Rings on the Globe (ShRG) and the Shadow of the Globe on the Rings (ShGR). It should be noted that the shadows are not always in the same place. Before opposition the ShGR will lie on the preceding side of the rings, vanish at opposition (i.e. the time when Saturn is opposite the Sun in the sky and due south at midnight) and reappear on the following side. It is always interesting to see how early one can spot the shadow's return after opposition. Similarly, the ShRG lies below the rings before opposition, vanishes at the time of opposition and reappears above the rings afterwards.

When viewing Saturn through a large telescope you will almost certainly see many different variations and delicate structures within the rings. David Gray's drawing in Figure 16 shows just how much detail can be seen in a large telescope on a good night.

Figure 16. An enormous amount of detail on Saturn and its rings as observed and drawn by David Gray with his 415 mm Dall-Kirkham.

Intensity Estimates. To make an intensity estimate for Saturn and its rings, first make a disk drawing as usual and then place your estimates on the features. Like those for Mars and Jupiter, the estimates should range from 0 (very bright white spots) to 10 (black sky). Shadows and the Cassini Division should be 10, the SEB is frequently 5 and so on.

During the course of an apparition, small white spots can suddenly appear in the various belts and zones. If you suspect a spot, you should get a CM timing; wait for the spot to reach the CM and record the time (in UT). You can then convert this into longitude and obtain a longitude estimate for the spot.

I shall say just one final thing about the satellites. Saturn has a large number (more than sixty) and no fewer than five are visible in a 4-inch telescope. Of them all, Titan is the largest and brightest. Although it is of no scientific use, I like to record the positions of the satellites and identify them when I go back indoors. It's also worth remembering while looking at Titan that a little piece of the UK is there on the surface – what a sobering thought!

This concludes my contribution on visual amateur astronomy. I do hope you give it a try, whether you're a newcomer to astronomy or someone who has been imaging for a long time. You really can't beat just looking – after all, isn't that how most of us got started?

FURTHER READING

Grego, Peter, *Solar System Observer's Guide*. This book provides a good introduction to Solar System observing and contains many tips for visual observers.

Sky at Night magazine. This magazine comes with a CCD which contains observing forms for planets visible each month with the correct tilt and phase.

The British Astronomical Association's website, www.britastro.org, is a good website for reviewing the latest lunar and planetary reports.

Stargazerslounge is a great forum to show off your latest images and get help and advice on all areas of amateur astronomy. Be sure to pay it a visit: stargazerslounge.com.

New Views of Mercury

DAVID A. ROTHERY

Upon reading a new translation of the works of Homer in 1816, the poet John Keats was moved to compose a sonnet about the experience, in which he wrote:

> *Then felt I like some watcher of the skies*
> *When a new planet swims into his ken;*
> *Or like stout Cortez when with eagle eyes*
> *He star'd at the Pacific – and all his men*
> *Look'd at each other with a wild surmise –*
> *Silent, upon a peak in Darien.*

Keats's 'new planet' metaphor may reflect either Herschel's sighting of Uranus in 1781, or the discoveries of the first four asteroids (1801–7), which were fresher in the memory and, to laymen such as Keats, also counted as 'new planets'. These days, new worlds are being discovered every month, in the Kuiper belt and as exoplanets around other stars. Call me jaded, if you like, but I no longer become excited by seeing an image of a new ice-ball as a single pixel, or by a hint of a hot Jupiter revealed by wobbles in a star's radial velocity curve. What does inspire me, though, is seeing a new planetary landscape for the first time. I suppose I am more analogous to Keats's explorer admiring a fresh vista than to a 'watcher of the skies' happening upon a new speck of light.

Those of us following events in 2008 shared a remarkable 'Cortez experience' when NASA's MESSENGER spacecraft revealed in great detail vast expanses of the surface of Mercury that had never yet been seen. This happened during the first two of MESSENGER's three planned flybys of Mercury, which were necessary to allow the probe to be captured into orbit about the planet in March 2011. MESSENGER's main job will begin then, but in the meantime the January and October 2008 flybys have revealed to us the Solar System's last great unseen expanses this side of Triton. None of us will live through the like again, though of course new close-up views of asteroids and Kuiper belt objects will provide revelations on a smaller scale; and more detailed imaging of parts of Mars, Titan, Enceladus and so on will continue to add to our understanding.

MARINER-10, MESSENGER AND MERCURY

In 1974–5, Mariner-10, the only other spacecraft to have visited Mercury, imaged about 45 per cent of its surface. Between them, the first two MESSENGER flybys imaged a further 51 per cent (Figure 1).

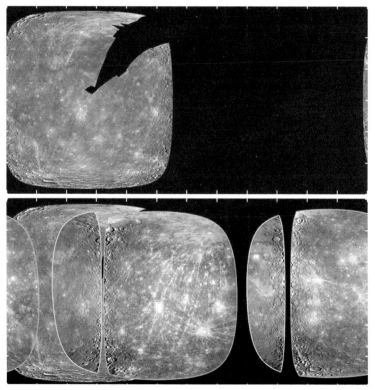

Figure 1. Mercator-projected image mosaics of Mercury showing Mariner-10 coverage (above), compared with the additional coverage after the first two MESSENGER flybys. The left-hand edge is 180° west and the right-hand edge is 180° east.

This amounts to 38 million square kilometres, which is an area bigger than Europe and North America combined. They also improved our knowledge of some regions previously seen by Mariner-10 by imaging them at higher resolution, under more suitable illumination conditions (Figure 2), and in colour. MESSENGER's colour information comes from imaging at eleven wavelengths in the optical part of the spectrum (400–1100 nm) and from a spectrometer called MASCS (Mercury Atmospheric and Surface Composition Spectrometer) which has poor spatial resolution but spans the spectrum from 200 nm (ultraviolet) to 1300 nm (near infrared) in high spectral resolution. The images illustrating this chapter are from MESSENGER's narrow-angle camera, which records in a single, panchromatic channel, at higher spatial resolution than the colour images (from orbit, pixels will be as small as 18 m across). MESSENGER also carries a magnetometer that

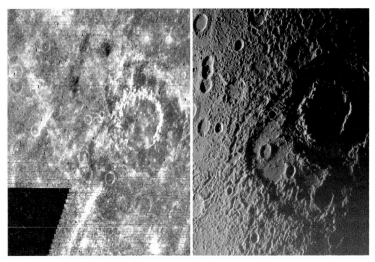

Figure 2. *Mariner-10* (left) and MESSENGER (right) views of the same area of Mercury. The 200-kilometre diameter double-ring crater Vivaldi is prominent on the right of each image. Mariner-10 saw this area with the Sun overhead (so albedo features are prominent, but topography is suppressed), whereas MESSENGER imaged it shortly before sunset (so topography is clearly seen). The comparison does Mariner-10 somewhat of an injustice, because low-Sun Mariner-10 images are almost as attractive as MESSENGER low-Sun images. (The missing data in the south-west corner of the Mariner-10 image does not match the data-gap shown in Figure 1.)

is vital for understanding Mercury's magnetic field. This has about 1 per cent of the strength of the Earth's magnetic field, but it was one of Mariner-10's most surprising discoveries because it is several orders of magnitude stronger than the paltry magnetic fields possessed by Venus and Mars. Other instruments such as a laser altimeter and an X-ray spectrometer will really only come into their own when the orbital phase of MESSENGER's mission begins.

The basic facts about Mercury are as follows. It is the Sun's innermost planet, having a semi-major axis of 0.387 AU and an orbital period of eighty-eight days. Mercury's rotation has become tidally locked to the Sun; not in synchronous rotation (i.e. one rotation per orbit, like the Moon's relationship to the Earth), but exactly three rotations per two orbits. This makes Mercury's day (sunrise to sunrise) twice as long as its year. If you find this hard to imagine, look at it this way: a planet that rotated once per orbit would have permanent day on one side and permanent night on the other. A planet needs to rotate *twice* per orbit to have *one* day per year. Mercury rotates more slowly than this – one-and-a-half rotations per orbit – so its day has to be longer than its year.

Soon after it was formed, Mercury is likely to have been spinning with a rotation period as short as eight hours, which was gradually slowed down because of tidal drag from the Sun. Why it settled into a 3:2 spin:orbit resonance rather than 1:1 is probably connected with the shape of its orbit, which is the most eccentric orbit of any planet (aphelion 0.47 AU, perihelion 0.31 AU). The 3:2 resonance is such that exactly opposite hemispheres face the Sun during alternate perihelion passages.

The subsolar point on the surface at perihelion reaches a temperature of nearly 430°C, whereas the coldest temperature (just before sunrise) is below –180°C. Nowhere else in the Solar System is known to experience such a wide diurnal temperature range. Being airless, Mercury's surface is unprotected from meteorite bombardment, and is bathed in solar radiation of all wavelengths. Its magnetic field shields much of the surface from the charged particles of the solar wind most of the time, but probably not always. Ions in the planet's exosphere, many of which may be escaping to space from the surface as a result of the harsh environment, are known to include hydrogen, helium and oxygen (discovered by Mariner-10's ultraviolet spectrometer), sodium, potassium and calcium (discovered with ground-based telescopes), and magnesium discovered by MESSENGER's MASC during the second flyby.

Another peculiarity of Mercury is its density of 5.43 tonnes per cubic metre. This is considerably greater than it would be if Mercury were of the same proportions of rock and metal as the Earth. The Earth, at 5.51 tonnes per cubic metre, is actually marginally denser than Mercury. However, Mercury is the least massive of the Sun's four terrestrial planets, at only 5.5 per cent Earth masses. Thus, its gravity is correspondingly weak and so it experiences much less internal 'self-compression' than the Earth does. We know from the way it reflects sunlight that Mercury's surface is rocky regolith, and the only reasonable way to explain the planet's overall high density (after taking the lack of self-compression into account) is that it must have a disproportionately large core. The cores of terrestrial planets are made largely of iron, which, at typical pressures, is about twice as dense as rock. In fact, since the days of Mariner-10 it has been surmised with reasonable confidence that Mercury's core must occupy about 42 per cent of its volume, corresponding to 75 per cent of its diameter. In contrast, Earth's core is 16 per cent of its total volume and 54 per cent of its diameter.

A question left unanswered by Mariner-10 was whether Mercury's magnetic field is generated by dynamo motion in a fluid outer zone of its electrically conducting core, in the same way as the Earth's magnetic field. The strength of the field seemed to indicate that this is likely, but factors arguing against included Mercury's slow rotation (hence slow stirring of any fluid part of its core) and the planet's small size leading to far greater loss of its original heat than that of the larger Earth, so that the core might now be entirely frozen. MESSENGER was designed to make various measurements from orbit to test for a partly fluid core, but the question seems already to have been settled by a remarkable study led by Jean-Luc Margot of Cornell University. Margot and colleagues used ground-based radar to measure deviations in Mercury's spin rate ('forced librations in longitude') during the course of several orbits, and in 2007 they announced that these amount to an oscillation of about 36 seconds of arc per orbit. That may seem tiny, but it is twice what would be possible if the planet were entirely solid. Margot and his colleagues concluded that the solid mantle of Mercury must be decoupled from the interior, so at least the outer zone of its core must be molten. MESSENGER laser altimeter measurements of Mercury's libration may strengthen this conclusion (and enable more reliable estimates of the core's size), but it seems unlikely that they will overturn it. A significant implication of a partly molten core inside such a small body is that there probably needs to be a few per cent of sulphur mixed with the iron to suppress the freezing temperature. However, so much sulphur does not fit well with the expected composition if Mercury was assembled from material condensing so close to the Sun as Mercury's present orbit, so one way or another there is still a mystery to be solved.

THE MESSENGER IMAGES

But what is revealed on the new images that so excited me? Figure 2 shows the additional detail visible on some of the MESSENGER images, and is a fairly typical view of heavily cratered terrain on Mercury. Except for the smooth, possibly lava-filled floors of some of the craters more than 40 kilometres across, the terrain seen here is rugged and littered with impact craters. However, the roughness of this particular area owes much to chains of secondary impact craters and other ejecta architecture radiating away from Vivaldi and the older basin(s) that Vivaldi largely obliterates. This area is not so close to saturation with primary impacts as the lunar highlands.

Everywhere on Mercury is cratered, but you don't have to look far to find areas that are considerably less dominated by craters than the neighbourhood of Vivaldi. Figure 3 is a good example, and I will describe this area at some length. Most of it, especially in the south, is relatively smooth. There are plenty of crisp-looking craters with prominent central peaks punched into the smooth surface, and which are clearly younger than it. However, if you look carefully you should soon realize that there are also at least as many craters that are hard to discern because they have been largely buried by the smooth plains. The rims of these craters can be traced, in whole or in part, but their floors are flat and approximately level with the surrounding plains. Evidently those craters were formed on an older surface, and then became draped by a younger plains-forming deposit that almost, but not quite, entirely buried the pre-existing craters.

On the basis of Mariner-10 images there were two main competing schools of thought about what could have formed the plains and buried

Figure 3. A mosaic of MESSENGER images showing a 1,300-kilometre wide area straddling Mercury's equator. The scarp named Beagle Rupes is in the upper left, and cuts across an elliptical crater named Sveinsdottir.

the pre-existing craters: volcanic lava flows or sheets of fragmentary ejecta distributed from the sites of major impacts. However, a consensus has emerged on the basis of the MESSENGER images that the plains are volcanic; thick series of lava flows spread across the terrain, overtopping the rims of craters and spilling into their interiors. After emplacement, the surface of a lava flow sinks slowly downwards because of escape of gases and collapse of internal void spaces, and also because of general thermal contraction. Subsidence is least where the lava is thinnest, allowing the buried rims of craters to re-express themselves subtly upon the surface topography. Craters that were buried but which can still be seen are known also on the Moon and Mars, where they are called 'ghost craters'.

In Figure 3, lavas have partly flooded the unnamed 250-kilometre diameter double-ring crater in the south-west. The terrain outside this crater is moderately smooth and I think it is essentially volcanic too, but it has more superimposed craters than the interior of the double ring, and so lava emplacement beyond the outer ring must have ended before the final flooding of the crater floor. However, if you trace the plains eastwards you pass down over a straight, sunlit scarp oriented from north-west to south-east. The terrain downhill of the scarp is smoother, and clearly the most recent lava emplacement on these low plains post-dates the formation of the scarp. The scarp is presumably a fault, and you can even see a 40-kilometre diameter impact crater pre-dating the fault whose north-east edge has been down-dropped by the fault, and then partly flooded by the lavas that ponded against the scarp.

There are some subtle 'wrinkle ridges' in the lava plains, hard to make out at this scale. These are symmetrical ridges formed during subsidence of the surface, and such features are well known on the lunar maria. On Figure 3, lobate scarps are much more prominent. These are a characteristic feature of Mercury first identified on Mariner-10 images, and they are steps in the terrain, across which the surface drops by up to about 1 kilometre. As their name suggests, they have sinuous or lobate traces, unlike the straight scarp that I have pointed out, which is something of an anomaly both because of its straightness and because it is older than the local smooth plains. Lobate scarps are longer than wrinkle ridges, and unlike them can cross from one terrain unit to another. The fact that lobate scarps cut the smooth plains shows that they must be younger than them. It is thought that they are the surface expressions of shallow-dipping thrust faults, and that the surface of the crust on the high side has been pushed over the low side. Since the days of Mariner-10 the favoured explanation for their origin is an episode of global contraction, caused by the cooling of and shrinking of the core more than 3 billion years ago. Adding up the contraction implied by all the known scarps leads to an estimate for the total reduction in Mercury's diameter of about 2 kilometres.

The formal term for these scarps, both singular and plural, is *rupes*, which is Latin for scarp. By convention, the *rupes* on Mercury are named after ships employed on voyages of discovery. It was fitting in view of the imminent bicentenary of Charles Darwin's birth that the most prominent *rupes* discovered by MESSENGER received the name Beagle Rupes, in honour of HMS *Beagle*, upon which Darwin sailed when he amassed the evidence in favour of evolution. Beagle Rupes is in the north-west quarter of Figure 3. There are a variety of reasons why I, as a geologist, regard Beagle Rupes as being of special interest.

One is that at either end it turns into straight faults at an angle to the thrust-front, across which the displacement must be partly sideways as opposed to purely compressional. Another is that traces of other scarps can be glimpsed in the poorly illuminated region between Beagle Rupes and the terminator. We must await imaging from orbit when the Sun is further west in the local sky to reveal how these relate to one another, so that we can unravel the tectonic history of the region.

Even on the Moon, where there is no doubt that the maria are filled by lava, the vents through which the lava was erupted are notoriously hard to find. However, there is at least one moderately convincing candidate for a volcanic vent on Mercury, proposed by Jim Head of Brown University, and colleagues (Figure 4). Stereo imaging and laser altimeter observations after MESSENGER has achieved orbit are awaited to confirm whether the vent craters are indeed at the summit of a broad dome. High-resolution imaging from orbit will probably be necessary before we can glimpse sinuous rilles, like the kilometre-wide

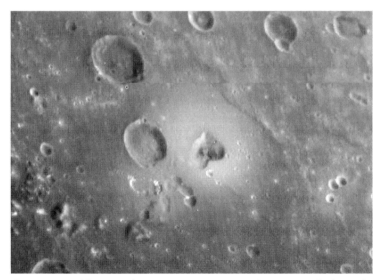

Figure 4. A 150-kilometre wide MESSENGER mosaic, centred on a proposed volcano. The main vent is the kidney-shaped structure in the centre, which appears to be at the summit of a broad rise. The 20-kilometre-diameter impact crater to its west appears to predate the growth of the volcano, because the east-facing part of its rampart has been almost buried below the pale material that forms the volcano.

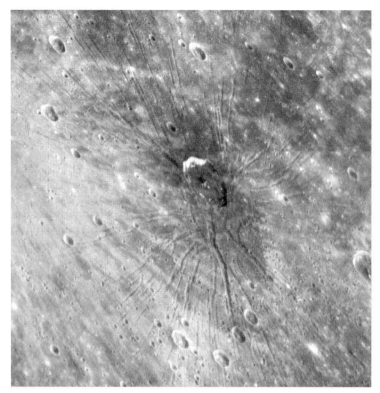

Figure 5. A 400-kilometre wide mosaic of two MESSENGER frames in the centre of the Caloris basin. The radial fractures, which have been named Pantheon Fossae, may reflect updoming of the central floor of the basin. The 40-kilometre impact crater (Apollodorus) in the centre is presumed to be a 'lucky strike' that happened to hit the bull's eye at a later date.

Hadley Rille on the Moon, visited by Apollo 15. These are collapsed tunnels or channels along which lava flowed. Mercury's surface gravity is twice that of the Moon (about one-third that of the Earth), so sinuous rilles on Mercury would be expected to be narrower than their lunar counterparts.

And what of Mercury's largest impact basins? Mariner-10 famously discovered the 1,600-kilometre diameter Caloris Basin, but saw only half because it straddled the terminator. MESSENGER saw the whole of it during its first flyby, revealing radial fractures in its centre (Figure 5), and concentric fractures inside its rim (Figure 6). It also

Figure 6. A 400-kilometre wide MESSENGER image of an area just north-west of Figure 5. Here the fracture pattern is concentric about the basin.

showed that the smooth plains fringing Caloris have a 40 per cent lower density of superimposed craters than the plains inside Caloris. Thus, the exterior plains have to be younger than the interior plains, so certainly cannot be composed of ejecta excavated during the Caloris impact. This is further proof that the plains are volcanic, and in fact Figure 4 is a close-up of part of those exterior plains.

Before the MESSENGER encounters, there had been speculation about the existence of a second major basin on Mercury, informally known as the Skinakas Basin. This was inferred from a low-albedo feature centred near the equator at 280°W (80°E) that had been 'seen' using computer-aided image synthesis, using frames obtained with a 1.29-m telescope at the Skinakas Astrophysical Observatory in Crete.

However, expectations faded when the first MESSENGER flyby saw no ejecta architecture relating to the basin, which ought to have extended into the region of Figure 3 from beyond the terminator. The second flyby, imaging the site of the basin in sunlight (Figure 7), saw no trace of it, and appears to have disproved its existence. However, there plenty of interesting features were revealed. For example, the close-up view in Figure 5 shows what appears to be a conventional lobate scarp, as a shadowed feature running north to south and crossing the most prominent crater. However, if you look closely there is an opposite-facing, sunlit scarp less than 30 kilometres to its west, and snaking roughly parallel to it, so that what we have here is a long thin strip of terrain pushed up by thrusting in two opposite directions.

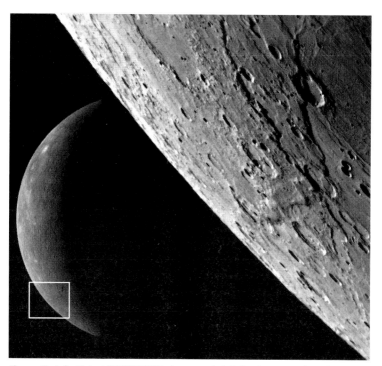

Figure 7. Left: Distant MESSENGER view recorded during its approach to the second flyby. The Skinakas Basin, if it existed, ought to straddle the equator just west of the terminator. Right: MESSENGER close-up view of the area outlined by the box on the left. The prominent crater near the upper right is 30 kilometres across.

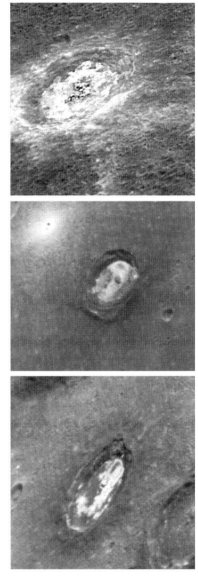

Figure 8. Three approximately 50-kilometre-diameter craters with mysterious bright deposits on their floors, seen by MESSENGER. Left: Sander, in the north-west of the Caloris Basin. Centre: Kertész, in the south-west of the Caloris Basin. The bright spot to its north-east is the fresh ejecta surrounding a small younger crater. Right: unnamed crater seen under high Sun during the second flyby.

Even more perplexing is the texture on the plains in the north-east corner of the image. Here, the low-angle solar illumination reveals a pattern of ridges and grooves, running north-west–south-east, but not quite parallel to one another. The texture puts me in mind of the grooved terrain on Ganymede, or (on a smaller scale) of the 'ball-of-string' (tectonic) terrain on Europa, but I would be very surprised if these share a formation mechanism in common. We may be seeing part of a generally radial fracture pattern on the floor of a large basin; not Skinakas, because this is too far south, and apparently lacking the outer zone of concentric fractures seen in Caloris.

Another peculiarity of Mercury revealed by MESSENGER is that a few of the impact craters are distinguished by having blotches of relatively bright (high albedo) material on their floors (Figure 8). These blotches are not symmetrically distributed, and show no indication of having been produced as part of the crater excavation process, but they could be pools of impact melt (now long-since frozen, of course). Alternatively, they may have formed later than the craters themselves, but how? Are they high-albedo lava flows? A few lunar craters have dark lava flows in patches on their floors, but (even discounting the brightness) these don't look very similar. Are they sites where some kind of relatively volatile material has exhaled through the crust and condensed at the surface? If the latter is correct, then are they stable, or can we expect to see changes over the coming months or years?

MESSENGER AND BEPICOLOMBO IN ORBIT

The answering of questions such as these can probably begin only with the orbital phase of MESSENGER's mission, which is intended to last for at least a year. The orbit will be eccentric. At its lowest, it will pass only 200 kilometres above the surface. This will always be over the Northern Hemisphere. The far point of the orbit, over the Southern Hemisphere, will be 15,000 kilometres away, which is a slightly greater range than for the closest of the flyby images seen in this chapter. MES-SENGER carries a relatively modest suite of instruments, but will be followed by BepiColombo, a joint project between the European and Japanese space agencies, which is due to arrive in orbit in 2020. Upon arrival, BepiColombo will divide into a Mercury Planetary Orbiter (MPO), in a relatively circular low orbit optimized for study of the surface, and a Mercury Magnetospheric Orbiter sharing the low point of its orbit with MPO but going much further out to explore the magnetosphere.

One of the greatest puzzles about Mercury is that its surface appears to have remarkably little iron. The iron oxide content of the surface silicates cannot be more than about 3 per cent, according to spectra obtained telescopically from Earth and by MASC during MESSEN-GER's first flyby. This would raise no eyebrows if Mercury's surface were made of anorthosite, like the lunar highlands, produced as a 'primary crust' by the flotation of low-density crystals from the original magma ocean. However, so much of Mercury seems to be lava flows (hiding the primary crust from view) that this low abundance of iron is very curious. Certainly it is far too low to be the kind of basalt that is well known on both the Earth and Moon. Average lunar mare basalts have about 20 per cent iron oxide, implying an iron oxide content in the lunar mantle of about 13 per cent. Even if lava flows on Mercury have as much as 3 per cent iron oxide, the implication is that Mercury's mantle could have only about 2 per cent iron oxide. This seems almost unthinkably low if Mercury grew from the same stuff as the Earth and the Moon (Mercury's enormous iron core is a separate issue).

There may be something wrong with the optical spectroscopy approach, which relies on detecting absorptions caused by the iron-oxygen bond. An X-ray spectrometer carried by MESSENGER, and a considerably more sophisticated (British-led) one carried by Bepi-Colombo, will detect fluorescent X-rays from atomic electron shells, and will detect iron (and several other elements) independent of chemical bonding. This will give us the total iron abundance in the crust, and show us how iron varies between different units. It is likely that some of the iron-oxygen bonds in Mercury's surface silicates have been broken (by 'space weathering') so that much of the iron now exists as sub-microscopic particles of metallic iron, which would not contribute to iron oxide absorptions in optical spectra. However, it is unlikely that this can be the main explanation for the apparent low iron abundance. Maybe the mineralogy is something unexpected, in which case BepiColombo's infrared spectrometer – a device lacking on MESSENGER and giving an independent way to determine surface mineralogy – may fill the crucial data gap.

I have high hopes that BepiColombo's X-ray and infrared spectrometers will reveal the composition not only of individual lava flows on Mercury, but also of the pale blotches in craters such as those in Figure 8, of the material thrown out as crater ejecta, and of the central peaks of craters, which have been uplifted from depth. Other instruments will analyse the dust, ions and neutral atoms that are continually leaking from the surface into space. We need to know all of this, and much more, before we can understand how Mercury was formed and how it subsequently evolved.

I conclude with a view from MESSENGER that sums up the rich diversity of this planet (Figure 9). It is a view towards the horizon (limb), of a kind unlikely to be recorded once the orbital phase of the mission has begun. In the foreground is rugged terrain with numerous craters. On the right there is a very clear flooded boundary where this rough area is overlain by smooth plains (lava), which extend all the way to the limb. The smooth plains flood numerous ghost craters, but are pockmarked by younger craters and further disfigured by lobate scarps and wrinkle ridges. To the left of centre is a fresh-looking 80-kilometre-diameter crater with a complex of central peaks (compare its pristine

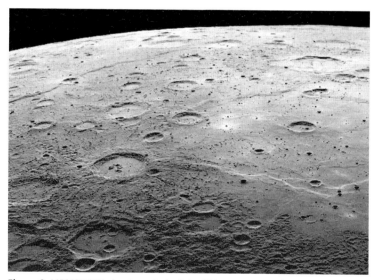

Figure 9. MESSENGER view towards the limb of Mercury, as seen during the approach to the second flyby. This covers the central part of the limb from the distant view in Figure 7, but has been rotated 90 degrees. The foreground is about 450 kilometres from side to side.

form with that of the ancient and degraded crater of similar size to its left). This must have been formed after the smooth plains, because it obscures the boundary between the smooth plains and the foreground rough terrain. A continuous ejecta blanket surrounds this crater for a distance of about one crater-diameter, beyond which chains of secondary pit-craters can be made out, radial to the main crater. I think this is a remarkable picture. Enjoy it, and look out for a similar view from a slightly different perspective during MESSENGER's third and final flyby on 29 September 2009.

Comet Halley's 1910 Return

MARTIN MOBBERLEY

Exactly one hundred years ago the most famous comet of all put on a spectacular show in the May night sky, in an apparition which excited, amazed and even terrified much of the Earth's population. Many readers of this *Yearbook* will remember the more recent return of comet 1P/Halley in 1985/86. If you were a keen amateur astronomer a quarter of a century ago (yes, it was that far back!) you will have indelible memories of that era. I would like to indulge in a few of mine before delving back a full century in time.

Breakfast TV was in its infancy in 1985 but, despite this, the BBC found time to feature amateur astronomers Alan Young and Ron Arbour on their early morning programme as soon as they secured the first UK photographs of Halley in August 1985. In a separate *Sky at Night* programme shortly afterwards; the one and only Patrick Moore featured Ron and his 16-inch (40cm) Newtonian patrolling for super-novae and bagging the comet too. December 1985 and early January 1986 saw UK amateurs struggling to photograph the comet (using insensitive stuff called film!) in mainly cloudy skies before it plunged south. Sadly, the comet was the last thing on astronomers' minds when the Space Shuttle exploded on 26 January, two weeks before Halley's 1986 perihelion. Ironically, the Challenger disaster occurred just two days after a brilliant NASA success: the Voyager 2 flyby of Uranus. Halley reached perihelion on 9 February 1986 and peaked in brightness in early March at a rather disappointing magnitude 2.5 in the southern sky. The Giotto spacecraft arrived at Halley on 13 March and sent back the first images of a cometary nucleus; technically the mission was a huge success, but the farcical live TV programme covering the rendezvous is something best forgotten: on the night nobody had a clue what the pictures were showing!

But all this is relatively recent history of almost a quarter-century ago; go back exactly one century to 1910 and Halley was a much brighter, awe-inspiring, even scary sight in the night sky. Frankly, in living memory, only one comet, Hyakutake (C/1996 B2), can stand comparison with Halley of 1910, as a comet with a sky-spanning tail stretching across a dark sky. Yes, there have been brighter comets, like McNaught's of January 2007 (C/2006 P1), but not in a truly dark sky, well away from the horizon, and with a tail spanning 100 degrees. With those attributes a 'Great Comet' can become quite a frightening sight. In addition, without stating the obvious, 1910 was a completely different world, with very little known about comets and the communication of information being painfully slow. Indeed, in July 1910 the murderer Dr Crippen was famously arrested because of a ground-breaking piece of new-fangled communication: a wireless telegraph sent from the SS *Montrose*, in the Atlantic Ocean, to Scotland Yard.

PRE-1910 RETURNS

Before we can fully appreciate the 1910 return of comet 1P/Halley it is important to realize just what it is that makes this comet so famous. Unfortunately we are living in an era in which the last return of Halley (1986) was rather poor and the circumstances of its next return (2061) even worse; in both cases, purely by bad luck, as the comet passes a long way from the Earth when it returns. But many returns of Halley have been quite spectacular. Comets can become awesome for a variety of reasons: they might be inherently very active or very big (like Hale-Bopp), they may fly close to the Earth (like Hyakutake or Iras-Araki-Alcock), or they may fly close to the Sun and brighten dramatically in twilight (like McNaught). To be honest, most 'Great' comets have more than one thing going for them. Halley is extra special because it is a very active comet that passes quite close to the Sun at perihelion (0.59 AU) and can also come dramatically close to the Earth and, above and beyond this, it returns every seventy-six years (or so). No other comet with such a healthy set of characteristics returns so regularly. All the other truly 'Great' comets have periods of thousands or tens of thousands of years, or longer. One might make an exception for 109P/Swift-Tuttle (period 130 years) or even 153P/Ikeya-Zhang, but that splendid comet, last seen in 2002, has a period averaging about 350 years. An entertaining rhyme summarizing Halley's characteristics, supposedly written by the tenth Astronomer Royal, Harold Spencer Jones (1890–1960), reads as follows:

Of all the comets in the sky,
There's none like Comet Halley.
We see it with the naked eye,
And periodi-cally.
The first to see it was not he,
But yet we call it Halley.
The notion that it would return,
Was his origi-nally.

Halley, like every comet, does not have a precise orbital period. Over time orbits evolve, because of the gravitational influence of the planets, especially Jupiter, and the non-gravitational factors, i.e. comets venting material. So, although 'seventy-six years' is often quoted as the orbital period for Halley, in practice anywhere between seventy-four and seventy-nine years is possible. The precise calendar date when the comet returns to perihelion (its closest point to the Sun) determines how close it will fly past the Earth and how awesome it will appear. Halley spends most of its life below the Earth's orbit. It rises through the Earth's orbital plane ninety-two days before perihelion and 270 million kilometres from the Sun. It then cuts above and inside the Earth's orbit, reaches perihelion and swoops back down and out again (Figure 1). Thirty days after perihelion Halley passes down through the Earth's

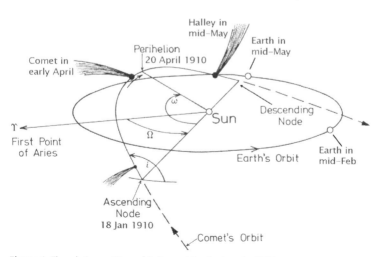

Figure 1. The relative positions of Halley and the Earth at the 1910 return.

orbital plane, 127 million kilometres from the Sun, and plunges below and outside the Earth's orbit. For the absolute best situation we need perihelion to occur from late February to late April with an Earth close approach, post-perihelion, from early April to late May, when Halley is plunging south. Pre-perihelion encounters with Halley, from late September to early October, can still be impressive (perihelion dates from late October to early November), but the comet is further away and has not heated up as dramatically as after perihelion. The perihelion date in 1986 was 9 February. If that date had been just two or three weeks later the return would have been spectacular in the extreme; as it turned out, those two or three weeks made all the difference between an awesome comet and a naked-eye smudge.

Some of the most memorable historic returns of Halley are listed in the table below; in addition, I have added the return of 1986, and the next two, in 2061 and 2134. It should be borne in mind that the dates up to and including 1378 are in the Julian calendar and by the return of Halley in that year the calendar was so out of step that the 10 November 1378 perihelion date is equivalent to 18 November in today's Gregorian calendar.

Year	Perihelion	Perigee	Dist. (AU)	Peak Mag.
141	Mar 22	Apr 22	0.17	−1
374	Feb 16	Apr 02	0.09	−1
607	Mar 15	Apr 19	0.09	−2
837	Feb 28	Apr 11	0.03	−3
1066	Mar 20	Apr 24	0.10	−1
1301	Oct 25	Sep 23	0.18	+1
1378	Nov 10	Oct 03	0.12	+1
1910	Apr 20	May 20	0.15	0
1986	Feb 09	Apr 11	0.42	+2.5
2061	July 28	July 29	0.48	+3?
2134	Mar 27	May 07	0.09	−1?

Table 1. A few selected returns of 1P/Halley from the last two millennia with perihelion dates, perigee dates of the closest Earth approaches, corresponding perigee distances to the Earth in AU (1 AU = 149.6 million kilometres) and the peak magnitude achieved/likely to be achieved. Dates up to and including AD 1378 are from the Julian calendar. The next two passes have also been included.

THE 1909 RECOVERY

Ever since Edmond Halley calculated (in 1705) the first predicted return of what would be regarded as *his* comet, much kudos has been associated with being the first astronomer to spot the incoming 'first periodic comet' (1P). Halley realized that the comets of 1531, 1607 and 1682 had almost identical orbits and were separated by intervals of roughly seventy-five years. He successfully predicted that the same comet would return in 1758, although, as it turned out, this would actually occur sixteen years after his death. That first predicted recovery of Halley, nowadays designated 1P/Halley, was made by Johann Georg Palitzsch, a German farmer and amateur astronomer. In the twenty-first century, when thousands of orbits can be calculated in seconds on a modern PC, it is impossible to imagine the mental torture that went into every nineteenth-, eighteenth- and seventeenth-century cometary orbit calculation (and even the measurements at the eye-piece), but it was a challenge that stretched the finest mathematicians over many months. Even after an orbit was calculated there was always considerable uncertainty as to precisely when and where a comet would re-emerge. Searching the skies for a returning comet was like a blind man searching in a coal cellar for a black cat that might not even be there! The 1910 return of Halley would be the first one where astronomers were waiting primed with photographic film, and so the keenest astrophotographers of the day were up for the challenge of being the first to spot it (Figure 2).

Halley had last been seen in 1835 and the American writer and humorist Mark Twain was born on 30 November of that year, just two weeks after the comet reached perihelion. In his biography, he famously said, 'I came in with Halley's comet in 1835. It's coming again next year [1910], and I expect to go out with it. It will be the greatest disappointment of my life if I don't go out with Halley's Comet. The Almighty has said no doubt, "Now here are these two unaccountable freaks; they came in together, they must go out together."' Twain would not be disappointed. He would die on 21 April 1910, the day following the comet's 1910 perihelion passage, but before its dramatic climax.

But, stepping back a year or two, with the extra power of photography, astronomers had started searching for Halley as early as 1908.

Figure 2. A famous Heath Robinson cartoon depicting astronomers at Greenwich employing all manner of optical devices to spot the incoming comet.

The two men with the best chance of recovering it photographically were, arguably, Max Wolf, at Königstuhl Observatory at Heidelberg, Germany, and the tireless astrophotographer Edward Emerson Barnard of Yerkes Observatory, near Chicago. Of course, astronomers needed to know where to search and there were surprisingly few serious attempts to determine the 1910 perihelion return date and corresponding ephemeris for Halley in the years building up to its return. In 1863 A.J. Ångstrom had devised a prediction method which seemed to favour a January 1913 perihelion date. A year later P.G. Le Doulcet had arrived at a date of 24 May 1910. More than forty years later, with no other predictions being made and Halley's return imminent, in 1907 and 1908 A.C.D. Crommelin and P.H. Cowell of the Royal Observatory, Greenwich, tackled the problem and arrived at a perihelion date of 8 April 1910. Finally, A.A. Ivanov, as late as 1909, predicted a perihelion date of 22 April 1910.

As it turned out, Max Wolf at Heidelberg recovered the comet on the night of 11/12 September 1909. According to Morehouse (one of Barnard's colleagues at Yerkes Observatory) Barnard appeared to be devastated and 'white-faced' when he heard the news. He had been searching for Halley since October 1908 but his old rival Dr Wolf had snatched it first. After Wolf's recovery of Halley it soon became apparent that Crommelin and Cowell, and Ivanov too, had all got the month right. In fact, the perihelion date would be 20 April 1910, astonishingly close to Ivanov's prediction.

Following Wolf's recovery of Halley some very faint images of the comet were found to have been recorded on plates exposed at the Royal Observatory, Greenwich, and the Khedivial Observatory at Helwan, Egypt, on 9/10 September and 24/25 August respectively. An earlier (29 August) plate by Wolf also contained a faint recording of the comet, as did a September plate by Barnard. With the comet's position in its orbit now pinned down it did not take long for the salient points of this apparition to be calculated. The comet would come closest to the Earth on 20 May, a month after its return to perihelion. It would be an excellent return, even if it would not be record-breaking. If the comet had returned slightly later, the apparition would rapidly have become an average one, but as it was Halley would pass within 23 million kilometres of the Earth on 20 May and be an awesome sight that month. Even in 1910 commercial organizations realized that they might use this cometary return to promote their wares and a Halley connection started to appear in a number of newspaper advertisements (Figure 3).

At recovery, Halley was a 16th-magnitude fuzz roughly 10 arc seconds in diameter, but it brightened rapidly. One month later the

Figure 3. Even in 1910 Halley was ruthlessly exploited in newspaper commercials, such as this Waterman pen advertisement.

comet was 14th magnitude and by mid-November it had brightened to 12th magnitude with a coma up to 25 arc seconds across. By Christmas 1909 Halley was a 10th-magnitude object and various coma sizes up to 3 arc minutes were being reported. As 1910 started the comet continued to brighten above 10th magnitude and attract publicity, but there was a surprise in store: another object was about to grab some headlines.

THE IMPOSTER

On 13 January 1910 workers at the Premier diamond mine in Cullinan, South Africa, saw a bright comet in the dawn sky. The local Johannesburg newspaper, *The Leader*, reported this as a sighting of comet Halley. However, this was not Halley at all but a completely new and impertinent comet trying to steal the show! In future years, it would simply be referred to as the Great January Comet, or the Daylight Comet of 1910.

In the 1980s, with Halley returning again, it was often stated that those octogenarians who claimed to have remembered Halley's 1910 return were actually confused by childhood memories of this bright imposter. While that may be the case for a very few claims, Halley would certainly not disappoint those who saw it in a dark sky four months later. It is far more likely that Halley would have been the comet that stuck in the memory.

Nevertheless, the Daylight Comet was briefly spectacular, but it arrived unannounced and peaked in a twilight sky. It reached perihelion just four days after discovery, on 17 January 1910, at only 0.129 AU from the Sun. Although there were a few reports around perihelion of the comet being seen just a few degrees from the Sun, in daylight, and brighter than Venus, most observers (including those in the UK who knew where to look) reported a peak in brightness between 1st and 3rd magnitude in the weeks after perihelion and with a tail between a paltry 1 degree and a stunning 50 degrees in length. The longest tail lengths were generally reported towards the end of January, with experienced observers reporting that the comet had faded to 3rd magnitude by that time. February saw a dramatic fade in the Daylight Comet's brightness, from around 5th magnitude on 10 February to 9th magnitude by mid-month. The Daylight Comet had certainly been spectacular, but only for the briefest of periods in the second half of January. If you had cloudy skies or did not know it was there, it would have passed you by.

HALLEY BRIGHTENS

Meanwhile Halley itself had continued its steady rise in brightness throughout January 1910, an election month in the UK with the Liberals returning under Asquith, with a hung parliament. Across the Channel in France, Paris suffered much flooding when the Seine overflowed. In Germany the man who had recovered the famous comet, Max Wolf, exposed a plate on 29 January showing a 2.5 arc minute-diameter coma and a tail some 20 arc minutes in length. The comet was a steady 9th-magnitude object by the start of February and in the first week of that month Halley's brightness pulled ahead of the rapidly fading upstart, the Daylight Comet. All the top astronomers of the day were now concentrating on comet Halley. Some legendary names from the early twentieth century were observing it every clear night, such as Georges van Biesbroeck (Uccle, Belgium), Knox-Shaw (Helwan, Egypt), Millosevich (Italy) and Innes (Transvaal Observatory, Johannesburg), as well as Max Wolf at Heidelberg and E.E. Barnard at Yerkes.

By early March this most famous comet of them all was heading rapidly into the evening twilight as an 8th-magnitude fuzz, with Earth and comet on opposite sides of the Sun and moving in opposite directions (Halley has a retrograde orbit). Around 9 March the twilight won the battle and, despite some reports that the comet was now a 7th-magnitude evening object, observers had to switch observing times to do battle with the dawn twilight and await Halley's return to the skies in the far less sociable early hours. By the time many observers next spotted the comet it would be a naked-eye object. The keenest observers recovered a much brighter Halley in the second week of April 1910, with estimates of its brightness ranging from 5th to 4th magnitude. Undoubtedly, from mid-April onwards, more people were observing Halley than at any time in its history. The comet passed perihelion, well above the Earth's orbital plane, on 20 April, but was already moving rapidly down towards the Earth's orbit. It was now a magnitude 3.0 object with a tail 2 degrees in length. The comet had entered the critical immediate post-perihelion phase, when the weeks of maximum heating from the Sun were finally soaking into the icy nucleus creating the most activity. In addition, the comet was now rapidly drawing closer to the Earth. The result of these two factors conspiring would be dramatic. By the first week in May Halley's comet reached magnitude 2.0 with a tail some 18 degrees in length – an impossible object to miss in a dark sky. At the next return, in 1986, Halley would peak in the southern sky, but the much later perihelion date in 1910 meant that Halley was well placed for Northern Hemisphere observers, despite the encroaching summer twilight. It was descending towards the Earth from above, keeping its declination nice and healthy, peaking at +20 degrees north on 19 May, just one day before its closest approach.

PARANOIA SETS IN

Astronomers had realized shortly after Halley was recovered, and its orbit refined, that the Earth would be precisely downstream of its long tail in mid-May. In addition, the pioneering spectroscopic work undertaken at Lick (with the 36-inch Crossley reflector) and other observatories had revealed the presence of poisonous cyanogen (cyanide gas) and carbon monoxide in Halley's tail. For some members of the public for whom comets were still regarded as portents of doom, the idea of a giant tail stuffed to the brim with cyanide flapping around the Earth was enough to trigger a panic attack (Figure 4).

The fact that the density of material in a comet's tail was more tenuous than the most perfect Earth laboratory vacuum did not seem to allay these people's fears and, as ever with human nature, one gullible man's fear quickly became another man's profit margin. The pedlars of wonder tonics and elixirs saw a chance to make a quick buck in 1910 and as Halley brightened dramatically a variety of life-saving options appeared on the market, including comet pills and cyanide-proof gas masks. The situation was not helped by newspapers advertising these wares and reporting cases of mass hysteria where city dwellers were plugging their doors and windows to block the poisonous vapours. Remarkably, the residents of the city of Chicago, not far from the famous Yerkes Observatory where E.E. Barnard worked, appeared to be sealing their windows and doors with the greatest vigour. By

Figure 4. A famous French cartoon of 1910 depicting the options available on 19 May, the 'end of the world', when Halley's tail would sweep over the Earth.

contrast, in other cities, residents were holding comet parties, and in New York sales of telescopes were reported to be outstripping the supply.

Such was the interest in Halley that the *New York Times* appointed a special reporter, Miss M. Proctor, to produce daily comet reports on Halley during its peak period in May. The comet was given top billing from 9 May to 22 May, when it was headline news. Headlines included the following:

23 April: Women and foreigners attribute darkness over Chicago to comet; some become hysterical.

9 May: Bermuda observers report comet acting strangely following King Edward's death. [Edward VII had died on 6 May]

14 May: New York City hotel roofs being used for comet parties.

17 May: Boston will sound fire alarm if comet visible.

19 May: HALLEY'S COMET BRUSHES EARTH WITH ITS TAIL; 350 American astronomers keep vigil; reactions of fear and prayer

repeated; all-night services held in many churches; 1881 dire prophecies recalled by comet scare.

HALLEY CROSSES THE SUN'S FACE

Although Halley had moved westwards from the evening twilight sky to the morning twilight sky, travelling through Pisces in March 1910 and passing less than 6 degrees north of the Sun on the 27th, the gradual coming together of the Earth and the comet near Halley's descending node would reverse this apparent westward trend around 25 April. The comet briefly clipped the Pisces/Pegasus border before hurtling back east, through Pisces. As May progressed, with the comet past perihelion and heading towards a rendezvous with the Earth, the comet's motion eastwards increased dramatically and the head rapidly started to depart the morning sky and approach the morning twilight once more (Figure 5).

However, with the comet now so close to the Earth its head would not be lost in between the morning and evening sky for long: it was moving too fast through the sky (up to 15 degrees per day, due east). But there was to be an extra twist to this comet–Earth flyby. As the comet plunged towards the Earth's orbital plane (the ecliptic) and its closest point to the Earth, it would be in a perfect straight line between Sun and Earth – so perfect that the comet's nucleus would appear to pass right across the face of the Sun for twenty-five minutes around 04:00 UT on 19 May and, in theory, Halley's long gas tail would indeed sweep across the Earth itself on that day. Remarkably, the comet was observed in strong twilight low in the 18 May morning sky, although there do not appear to be any cast-iron observations of it by expert observers in the equally strong twilight of the 19 May evening sky. However, there are indications of it being visible from Tasmania in daylight, near the Sun, on 19 May. It should be borne in mind that the comet's tail (or tails) was so absurdly long at this time, because of its proximity to the Earth, that even though the head was immersed in the solar glare, the tail was not, except at that critical time when the comet passed directly over the Sun. The straight gas tail and curved dust tail were visible in both the morning and evening skies for much of mid-May. Of course, many leading professional astronomers with suitable equipment were monitoring the Sun to see if the head of

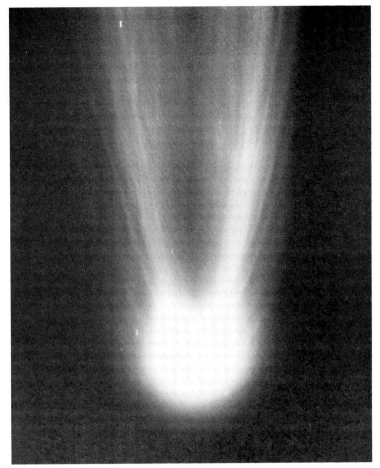

Figure 5. A famous photograph of Halley taken a day before New Moon, on 8 May 1910. (Image courtesy of Carnegie Institute of Washington Observatories.)

Halley could be seen crossing its brilliant face. However, observers at more than a dozen of the world's finest observatories across the globe failed to witness any transit of the nucleus across the Sun's blindingly bright disc. (As we now know that Halley's nucleus is $16 \times 8 \times 8$ kilometres it would have appeared, at 0.15 AU from Earth, only as a 0.15×0.07 arc-second dot crossing the Sun.) Nevertheless, both Max Wolf, Halley's recoverer, and E.E. Barnard reported unusual daylight atmospheric phenomena (solar haloes, intense twilights and prismatic effects) on the same day as the Earth was predicted to pass through the comet's gas tail, implying that tail particles in the Earth's vicinity were dramatically scattering the sunlight arriving at the Earth. On the days before and after Halley's closest approach there was much public interest in the potential Earth 'collision' with Halley's tail. At Treptow Observatory in Berlin, and at other observatories, huge crowds gathered on the night of 18/19 May in the hope of seeing the tail and hearing the latest news.

AN EXTRAORDINARY TAIL

By far the most amazing feature of Comet Halley at the 1910 apparition was the sheer length of its tail, or tails. Not until eighty-six years later and the close flyby of Comet Hyakutake in late March of 1996 would a similar phenomenon be observed. The tails of Great Comets can easily stretch for tens of millions of kilometres. When one passes within, say, 25 million kilometres of the Earth, so that the tail stretches right past us, and is in a dark sky, the sight can be extraordinary. In 1996 I was fortunate enough to be in Tenerife for the close fly-past of Hyakutake, which actually passed slightly closer to the Earth than Halley and also reached 0 magnitude. With direct vision I could see the tail stretching 25 degrees. With averted vision it stretched more than 60 degrees, but working out where the tail faded into the background sky was almost impossible. Experienced observers at even darker sites reported much longer lengths for Hyakutake. Observers of Halley in 1910 witnessed a similar, but even more dramatic spectacle. The famous comet was a month past perihelion when it came closest to Earth, whereas Hyakutake would be inward bound. The Halley tail lengths reported by the most reputable observers in 1910 started to become impressive around the New Moon period of 9 May, when lengths of 25 degrees were reported. On the 13th Max Wolf reported a tail length of 52 degrees with a width of 3 degrees at a time when naked-eye estimates of the comet's brightness were between magnitude 0 and 2. On the 18th E.E. Barnard reported a tail spanning a colossal 107 degrees, just prior to the solar transit period, and the tail engulfing the Earth. After 20 May, with the comet's head once again visible in the evening sky, the reported tail lengths became even more surreal, despite the full phase of the Moon. Experienced observers like Knox-Shaw were reporting a staggering 140 degrees of tail, up to 15 degrees wide at the widest point. But five days later Barnard reported a tail length of only 25 degrees, which the comet maintained until the end of the month, with its magnitude dropping from 0 to 3. By late May Halley was fading noticeably as, in addition to being well past perihelion, its distance from Earth was increasing rapidly (Figure 6).

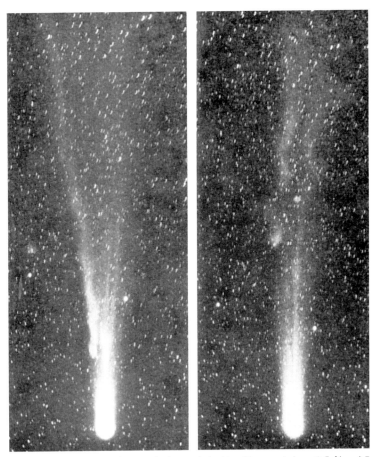

Figure 6. Comet Halley, photographed around the New Moon period, on 6 (left) and 7 (right) June 1910. Note that between the two dates a tail disconnection event has occurred and the old severed tail is floating away. (Images courtesy of Lick Observatory.)

Halley remained a naked-eye object until mid-June, but after that observers followed it with telescopes and by using photography. By mid-August 1910 Halley was once more disappearing into the twilight and when, eventually, it re-emerged in November 1910 it was no brighter than 12th magnitude. But the photographers persevered, as none of them would be around for the 1985/86 return. The last known successful observation of Halley was on 16 June 1911, when a forty-minute exposure by Curtis using the 36-inch Crossley reflector at Lick Observatory recorded 'a small hazy patch of utmost faintness'. Halley would not be seen again until 16 October 1982, when David Jewitt and Edward Danielson recorded a 24th-magnitude smudge in Halley's predicted position using an early CCD detector on the Palomar 200-inch (5-metre) reflector.

FUTURE PROSPECTS

I will be 103 years old when Halley next comes to perihelion in July 2061. That will be the most dismal return imaginable, with Halley unlikely to do better than 3rd magnitude and being very poorly placed in the sky. However, if current predictions are correct, the 2134 return will be a splendid one, with the comet even closer to the Earth than it was a hundred years ago in 1910. Indeed, the 2134 return will be very similar to the 1066 return, when the appearance of Halley was depicted on the Bayeux tapestry and linked to the Battle of Hastings and the downfall of King Harold. As I am unlikely to live to 176 years old I am hoping someone will remember to thaw me out of cryogenic storage . . .

The Anglo–Australian Astronomer

FRED WATSON

I wonder if the name Ben Gascoigne means anything to you? Maybe not. But if you are one of hundreds of British astronomers who have made their way to a fairly remote part of rural New South Wales over the past thirty-six years to observe with the 3.9-metre Anglo–Australian Telescope (AAT), you might well have come across Ben's name.

A rather elegant plaque, made of perspex and engraved 'Gascoigne's Leap' in large letters, is fixed to the inner walkway of the AAT dome at a point where it soars some 6 metres above the observing floor below. If you think this is starting to sound ominous, you're right, but it's not as ominous as it would have been had the plaque been fixed to the dome's outer walkway – the famed 'catwalk', where observers go outside not to model the latest in astronomical chic but to check the weather. From the catwalk, there is a 30-metre sheer drop down the side of the building to the cement below – and no one would survive the fall.

So what's the story of Gascoigne's Leap? Back in 1974, when the Anglo–Australian Telescope was brand new, and the building almost finished (but not quite . . .), Ben Gascoigne was the senior commissioning astronomer. First light had taken place on 27 April 1974, when he had obtained Test Photograph No.1 during the telescope's set-up procedure. (Incidentally, it comes as no surprise to today's AAT technicians that the adjustments made as a result of that photograph were carried out in the wrong direction and had to be repeated later – Murphy's Law was alive and well back in those days.)

A few nights afterwards, Ben was taking more photographs with the telescope when he had to pop out to check if there was cloud around. Coming in from the outer catwalk, he headed for the narrow gap in the inner walkway that would take him back to the telescope's control room. Unfortunately, in the dark, he lost track of exactly which of the dome's four outside doors he had entered, and when he walked through the gap he stepped not on to the control room walkway but into thin air . . .

He was *very* lucky. He managed to miss the protruding steel supports of the telescope's mirror trolley, parked under the walkway, by inches, and walked away with only a badly bruised elbow. Also, as luck would have it, he missed the steelwork of the walkway's inner safety rail, designed to prevent just such an accident, which was waiting on the dome floor to be installed. Needless to say, fitting it became the number-one priority for the engineering crew the next day, and safety in this hazardous building was never taken for granted again.

In many ways, Professor Ben Gascoigne, AO, epitomizes the resourceful, resilient and fairly unbreakable Anglo–Australian astronomer – even though the institution he worked for was actually the Australian National University (ANU). Ben brought to his work a strong background in optics, and an equally strong background in astronomy, concentrating on the evolution of stars and the development of the cosmic distance scale. Moreover, while he was neither Anglo nor Australian, hailing from New Zealand, he always embraced the community of British and Australian astronomers with warmth and generosity. Today, Ben is well into his nineties, and lives in retirement in Canberra, though he is still active in historical research. It's probably not surprising that he is held in unique affection within Australian scientific circles.

If there really is such a thing as an Anglo–Australian astronomer, it is perhaps best defined as one who has contributed to the well-being of the Anglo–Australian Observatory – and Ben certainly qualifies on that front. But there have been many, many others, and we will meet some of them in this article, as we celebrate three dozen years of AAT operations. Today, however, there is a new question to be asked. What will happen to the breed beyond 30 June 2010? For on that date, the Anglo–Australian Observatory will cease to exist in its present form, because the binational arrangement that has sustained it so far will finally come to an end . . . But let's begin at the beginning.

ANGLO–AUSTRALIAN COLLABORATION

As long ago as 1818, European explorers of inland New South Wales set eyes on a 'most stupendous range of mountains, lifting their blue heads above the horizon'. The words of their discoverer, John Oxley, are spiced with awe, and perhaps it was a similar awe that led him to honour a bigwig in His Majesty's Treasury by naming them 'Arbuthnot's Range'. Fortunately, this ridiculously inappropriate name didn't last. The range had been home to the Kamilaroi people for thousands of years, and their Aboriginal name soon reasserted itself. It is startlingly apt: Warrumbungle simply means 'crooked mountains'.

It was a group of astronomers from the ANU's Mount Stromlo Observatory in Canberra who first looked at the Warrumbungle mountains as a possible location for a major observatory. Their quest was for a new observatory site free from the growing light pollution of their home base. In 1964, the first telescope was built on Siding Spring Mountain, the 1,200-metre-high ridge in the Warrumbungles that is now home to half-a-dozen major astronomical facilities. That first pioneer was a 40-inch (1-metre) Boller & Chivens reflector that was made famous by the husband-and-wife team of Bart and Priscilla Bok while carrying out their studies of the Milky Way. It was followed by two smaller ANU telescopes and, perhaps more importantly, by infrastructure such as observer accommodation, power, water and a paved road to the country town of Coonabarabran, some 30 kilometres away. Like the mountains themselves, Coonabarabran has a Kamilaroi name – and it is no less apt: it means 'inquisitive person'.

When the British and Australian governments looked jointly at possible sites for a new 150-inch (or, in today's parlance, 4-metre) class telescope in the south during the late 1960s, the existing infrastructure at Siding Spring was certainly a consideration. But of prime importance was the quality of the atmosphere there, and tests revealed that it was, indeed, a suitable location for a new, large telescope. With spectroscopic conditions (i.e. clear apart from occasional thin cloud) prevailing for 65 per cent of nights, and photometric conditions (i.e. completely clear) for up to 50 per cent, together with reasonable conditions of atmospheric turbulence, Siding Spring was considered to be the best place in Australia for an optical observatory.

More recent site testing elsewhere in Australia has demonstrated that this is indeed true; but on a world scale, the continent lacks the geographical features necessary to produce the consistent high transparency and exquisite imaging of Hawaii or Chile. Thus, the AAT will probably remain Australia's largest telescope indefinitely, while Australian astronomers and their British counterparts invest in international collaborations (like Gemini) that establish 8-metre class (and larger) facilities on top of the extinct Mauna Kea volcano in Hawaii or in the high, arid Atacama desert of northern Chile.

What Siding Spring could boast of in the 1960s, and still can today, is that it is one of the darkest observatory sites in the world. Its geographical location results in a low natural sky brightness, but it also has the edge in minimal artificial background glow. In the early days, that was due simply to the observatory's remoteness from the metropolitan centres of Sydney and Newcastle. Now, with improved communications, growing inland cities and a decentralization of the residential population of Coonabarabran, much of that sense of remoteness has gone, and the observatory has to work hard in supporting environmental lighting legislation to protect its dark skies.

The construction of the AAT took place throughout the late 1960s and early 1970s (Figure 1). The telescope was intended to be modelled on the 4-metre Mayall Telescope at Kitt Peak National Observatory in the USA. In the event, however, significant changes were made to the design, although the basic equatorial horseshoe structure of the mounting was retained (Figure 2). In both instruments, the horseshoe diameter is 12 metres.

The AAT's 16-tonne primary mirror was cast from an ultra-low-expansion glass-ceramic material called Cervit by Owens-Illinois, Inc., in April 1969, and delivered later that year to Newcastle-upon-Tyne in England. Here, the firm of Sir Howard Grubb, Parsons & Co. Ltd began the long process of polishing the blank to turn it into a finished mirror under the direction of the late David Brown, one of Britain's most gifted optical scientists. In March 1973, it was declared ready for final testing, and its superb optical quality was revealed. If you imagine the 3.9-metre-diameter mirror expanded to the size of the British Isles, then the biggest departure from a perfect surface would be about the height of a pencil laid on its side. Not bad.

Eventually, having satisfied all the requirements, the mirror was transported to Australia and given a hero's welcome in Coonabarabran on 5 December. The low-loader on which it had been carried (rather

Figure 1. Taken in about 1970, this photo shows the massive bulk of the AAT building rising above the primeval landscape of the Warrumbungle Mountains in north-western New South Wales. (Photographer unknown; courtesy of David Malin Images.)

appropriately) from the New South Wales port of Newcastle was made to perform not one but two laps of honour around the small country town (Figure 3).

By then, the dome and building (Figure 4) had been finished for almost twelve months, allowing work to proceed on installing the telescope's mounting and control systems, which were built by the Mitsubishi Electric Corporation of Japan. Commissioning of the telescope took place throughout 1974, and by the end of the year useful scientific results were being obtained. It was also becoming obvious that because the AAT was to be operated completely under computer control (it was the first large telescope to do so), it was going to be a very fine instrument indeed, with a pointing accuracy of better than two seconds of arc – the angle subtended by a one-pound coin at a distance of 1½ miles. That was nothing short of astonishing in 1974, when one *minute* of arc was considered the norm. The telescope finally entered service as a scheduled, common-user instrument in June 1975.

Two other dates are of importance in the telescope's prehistory. On 22 February 1971, the Anglo–Australian Telescope Agreement came into effect and the Anglo–Australian Telescope Board was created. This

Figure 2. Although there are significant differences between the two telescopes, the basic equatorial horseshoe structure of the mounting for the 4-m Mayall Telescope at Kitt Peak National Observatory was retained in the AAT. In both instruments, the horseshoe diameter is 12 metres. (Image courtesy of Barnaby Norris.)

Figure 3. Fresh from its double lap of honour around the neighbouring town of Coonabarabran, the 3.9-metre mirror for the AAT tackles the final incline to the mountain-top in December 1973. (Photographer unknown; courtesy of David Malin Images.)

Figure 4. Comparable in size with a city office-block, the AAT building dominates the summit of Siding Spring Mountain, and is clearly visible for miles around. (Image courtesy of Ben Wrigley.)

binational board, with its three British and three Australian members, oversaw the strategic policy of the Anglo–Australian Observatory through the executive power vested in its Director. The underlying principle of the agreement was that the two governments would share equally the costs of construction, operation and maintenance, while astronomers from the two countries would have equal shares of observing time.

This resolved what for some time had been a political hot potato, the mode of operation being the subject of some disagreement between members of the British and Australian governments. It eventually allowed national differences to be put aside, and cleared the way for the other Big Day: the inauguration of the telescope by HRH Prince Charles on 16 October 1974.

HI-TECH IN THE BUSH

A place so generously endowed with natural beauty as the Warrumbungles attracts a steady stream of holidaymakers. They come to experience the flora and fauna, the spectacular scenery, the peace and quiet. If they camp in the Warrumbungle National Park, they might also experience – sometimes for the first time – a truly dark sky.

No visitor to the area can fail to notice the domes on the mountain top, particularly the insistent presence of the AAT. To a few it's an eyesore, but to most it's a monument to scientific endeavour. Nearly all of them, however, make the drive to Siding Spring to have a look.

The first thing they learn as they wind their way up the steep access road is that the AAT dome is the focus of a new 'virtual tour' of the Solar System, scaled so that the 37-metre diameter of the dome represents the Sun. On the mountain road, visitors pass scale models of Earth, Venus and Mercury, surprising in their smallness. If they take the drive into town, they'll pass Mars and giant Jupiter, but the other planets are strung out along the major highways of north-western New South Wales, each mounted on a large board replete with amazing facts they probably didn't know before.

Eventually, visitors will find themselves in a lift taking them to the observing floor of the AAT. It is one of only three lifts in the entire Coonabarabran district; the other two are a few metres away in the same building. The visitors are finally ushered into a long room, one wall of which is windowed from end to end and faces into the dome.

'Jeez! That is bloody *big*!' The reaction is always the same. Utter astonishment. Even though the AAT no longer ranks among the three largest optical telescopes in the world, as it did when it was built, it is still an imposing instrument. But it's not just the telescope that elicits

Figure 5. The cavernous void of the AAT dome is exaggerated in this wide-angle view, but it remains one of the largest telescope enclosures in the world. (Image courtesy of Ben Wrigley.)

surprise from onlookers. Because it was built at a time when science funding was more generous than it is today, the AAT boasts one of the biggest domes of any telescope in the world, and from the inside it looks huge (Figure 5).

The cavernous proportions of the dome have been a subtle contributor to the telescope's continuing front-line rôle in astronomy. When planning began almost two decades ago for a flagship auxiliary instrument to take the AAT into the new millennium, there was ample room to accommodate the necessary extension to the telescope's top end (Figure 6). That flagship instrument – the 2dF (for two-degree field) – was commissioned in the late 1990s and, as we shall see, set the direction for the telescope's future exploitation.

In the AAT's early years, it had to be all things to all users. It was by far the biggest telescope available to British and Australian astronomers and, then as now, they had to apply competitively for observing time through an independent time-allocation committee. For its part, the AAT had to fulfil all their requirements for imaging, photometry (measurements of brightness), spectroscopy (deciphering the barcode of information in starlight) and other techniques. Indeed, it had been built with versatility in mind, a set of three interchangeable top ends for the telescope providing a wide variety of focal ratios and focal stations. The auxiliary suite of more than a dozen instruments that was built to take advantage of this ranged from a large spectrograph (built by instrument scientists at the Royal Greenwich Observatory, then in Sussex) to advanced fibre-optic instrumentation built at the AAT's own laboratory in the Sydney suburb of Epping.

The telescope was originally designed with the idea that photographic plates would be the main detectors, both for taking images of the sky and for recording the rainbow spectra of celestial objects. For large-format astronomical imaging, that indeed turned out to be the case, and one of the unexpected contributors to the telescope's early reputation was the profusion of remarkable astronomical images made by its photographic specialist, David Malin (Figure 7). When, in the late 1970s, David pioneered a technique for recording celestial objects in true colour, the impact was dramatic, rocketing both the telescope and himself to world fame.

For spectroscopy, however, new electronic techniques were emerging, and the mid-1970s saw the installation of two ground-breaking detectors in rapid succession. First came the Image Dissector Scanner

Figure 6. One reason for the AAT's versatility is its system of interchangeable top-ends, allowing the telescope to be speedily reconfigured for different observing tasks. Here, the AAOmega top-end is removed after a spectroscopic observing campaign. (Image courtesy of Barnaby Norris.)

Figure 7. The beautiful spiral galaxy NGC 6744 in the southern constellation of Pavo (the Peacock) provides an irresistible target for the pioneering wizardry of David Malin. In this case, David used a large-format CCD camera with colleagues Steve Lee and Chris Tinney, rather than photography. (Image courtesy of Anglo-Australian Observatory/David Malin Images.)

(IDS), a brainchild of the observatory's first Director, Joe Wampler, and his collaborator, Lloyd Robinson. Within two years the IDS had been superseded by an even more powerful device in the shape of the Image Photon Counting System (IPCS), developed at University College London by Alec Boksenberg and his team – who were always known at the telescope as 'Boksenberg's Flying Circus'. When used with a spectrograph, the IPCS rendered the AAT at least as powerful as any other telescope in the world, providing astronomers with a unique tool for investigating the detailed properties of every kind of celestial object. Those detectors were later supplanted by the charge-coupled devices (CCDs) universally used today in both professional and amateur astronomy.

Two other circumstances conspired to increase the AAT's potency as a discovery machine. The first was that the southern sky was essentially unexplored by large telescopes. Even such obvious targets as the centre of the Milky Way Galaxy, our nearest neighbour galaxies (the Magellanic Clouds) and the closest globular clusters to the Sun had been observed only at low elevations by Northern-Hemisphere instruments.

The second was that the challenge presented by this virgin territory had been met by the British when they decided, in 1970, to go ahead with the construction of a wide-angle photographic survey instrument, the 1.2-metre UK Schmidt Telescope (UKST) (Figures 8 and 9). That instrument was formally opened at Siding Spring on 17 August 1973, and entered service some two weeks later. Its initial task was to photograph the whole of the southern sky not covered by its near-twin on Palomar Mountain in the USA during the 1950s, a job that eventually took the better part of a decade.

Figure 8. About half a kilometre from the AAT is the elegant building of its smaller sibling, the UK Schmidt Telescope, seen here at the start of a night's work. The telescope was built in 1973, and produced benchmark photographic atlases of the southern sky before being devoted exclusively to fibre-optic spectroscopic surveys. (Image courtesy of Ben Wrigley.)

Figure 9. The UK Schmidt Telescope is currently engaged in the RAVE survey of a million star velocities to probe the history of our Galaxy. To the left of the telescope's solid tube can be seen the robot enclosure for the 6dF fibre-optics system, while at lower right is the door of the spectrograph room. The large screen is not for showing midnight movies, but for calibration purposes. (Image courtesy of Ben Wrigley.)

Many reputations were made during that frenzied period when astronomers at the UKST collaborated with their counterparts at the AAT to scout out the most interesting southern objects and follow them up spectroscopically or photographically with the larger telescope. Eventually, in June 1988, the symbiotic relationship between the two telescopes was formalized when the UKST became part of the Anglo–Australian Observatory instead of an outstation of the Royal Observatory, Edinburgh. The UKST went on to complete its various photographic surveys, but from 2001 took on a new rôle as a spectroscopic survey telescope using a robotic instrument called 6dF (for six-degree field, by analogy with the AAT's 2dF). It is now engaged in RAVE (RAdial Velocity Experiment), a multinational project to measure the speeds and characteristics of a million stars.

From the beginning, scientists and engineers at the Anglo–Australian Observatory showed themselves to be adept at building novel instruments for use with the telescope. For example, the use of fibre optics in astronomy (an essential ingredient of both 2dF and 6dF), while not invented at the AAO, was transformed from an interesting novelty into a most effective technique at both the AAT and the UKST during the early 1980s.

Observations at infrared (redder-than-red) wavelengths became possible in 1979 with the introduction of the Infrared Photometer-Spectrometer (IRPS). The AAT could now see through dust clouds and study the earliest stages of star formation. A decade or so later, IRIS (infrared imaging spectrometer) was commissioned and had a major impact on the way the AAT was used. Both these instruments produced results that were spectacular in their day, including Solar System observations (such as the surface topography of Venus and the 1994 impacts of comet Shoemaker-Levy 9 fragments with Jupiter), detailed observations of galactic targets such as the Orion nebula and the Galactic Centre, and extragalactic observations.

Quite early in its history (and largely because of the effectiveness of the AAT), the observatory acquired a reputation as a 'finishing school' for young astronomers. Many of today's best-known figures in the science are products of that era. Because of his unique capabilities, one young astronomer was allowed to remain permanently on the staff rather than moving on after three years, as the rules demanded. His name was David Allen, and he was a person of extraordinary talent. As well as having an amazing breadth of astronomical expertise, covering every branch of the science from Solar System studies to cosmology, David was the driving force behind the AAT's many early successes in infrared astronomy. But his achievements didn't stop at astronomical research. He was also an accomplished science communicator, using broadcast and print media to get the message out that science is fun. As a youngster in England, David had been a protégé of the senior editor of this *Yearbook*, and became a regular contributor for many years. Indeed, the precedent became something of a tradition at the AAO, which has since produced a succession of outstanding science communicators. David remained at the observatory for nineteen years until his untimely death from a brain tumour in 1994.

MOST PRODUCTIVE IN THE WORLD

When the AAT celebrated its twenty-fifth birthday in 1999, it was with the recognition that the new millennium would bring challenges to a telescope that was starting to look small by world standards. More than a dozen ground-based telescopes with mirrors bigger than 6.5 metres in diameter were under construction or planned, and the Hubble space telescope had been producing breathtaking razor-sharp colour images of the Universe for more than half a decade. Moreover, even then, astronomers had their sights on a new generation of 'extremely large' telescopes with mirrors bigger than 20 metres in diameter.

The Anglo–Australian Observatory (AAO) had a proven track record in building innovative instrumentation, and already had an External Projects group to make its expertise available on a commercial basis in collaboration with other Australian or British institutions. Its customers included the European Southern Observatory's Very Large Telescope (VLT, the largest optical telescope in the world, comprising four 8.2-metre-unit telescopes), Subaru (the Japanese National 8.2-metre telescope) and the two 8.1-metre telescopes of the international Gemini project. Such interaction tends to be a two-way process, for involvement with these large telescopes feeds back into the scientific and engineering well-being of the AAO.

Recognizing this expertise, and embracing the scientific niches in which a 4-metre-class telescope on a less-than-perfect observing site could flourish, the observatory's management charted a future for the telescope that would keep it more than competitive in a VLT world. Specialization in a small number of world-class instruments rather than a large suite was a key ingredient, together with an emphasis on survey astronomy – the gathering of census-style data on large populations of celestial objects.

The flagship instrument then newly commissioned, 2dF, was unique, allowing the spectra of no fewer than 400 objects to be obtained simultaneously by means of robotically positioned optical fibres. Its first task was a three-dimensional survey of the distribution of galaxies within two-and-a-half billion light years of our own to provide a detailed cross-section of the Universe. That project, the 2dF Galaxy Redshift Survey, measured 220,000 galaxies and was completed in 2002, quickly becoming the richest source of AAO scientific papers to date. In 2005, it was used to find the 'missing link' between the ripples in the baby Universe some 13.7 billion years ago and today's distribution of galaxies. This remarkable work highlighted the importance of large-scale astronomical surveys, and confirmed that survey-type astronomy would continue to be a vital string to the AAT's bow.

Today, 2dF has metamorphosed into AAOmega. (The name is an in-joke among optical engineers, the so-called 'A-Omega' product being a measure of efficiency in an instrument.) AAOmega (Figures 10 and 11) utilizes 400 optical fibres like 2dF, but feeds them to a new, highly efficient and highly stable spectrograph. Since its completion in 2006, AAOmega has again been used for surveys of distant galaxies, but also for surveys of stars in our own Milky Way Galaxy. AAOmega will remain the world's most powerful spectroscopic survey instrument for a few years to come, but a further metamorphosis that will occur during 2010–11 is the addition of HERMES. This is a spectrograph especially designed for 'galactic archaeology' – the study of the evolution of our galaxy by investigations of very large numbers of its individual stars.

While AAOmega and HERMES represent the AAO's main arsenal of super-efficient multi-object spectrographs, the tradition of pioneering work in the infrared waveband has continued. Today's flagship instrument in infrared observation is IRIS2, a hybrid spectrometer-imager that was completed in 2003, and allows front-line astronomical research to continue through the Full Moon period, when the sky is too

Figure 10. Photon's-eye view of the AAT some 40 billionths of a second before it hits the telescope's 3.9-metre diameter mirror (left) and is reflected back up to the AAOmega fibres. The caterpillar cable tracks belong to the AAOmega fibre-positioning robot. (Image courtesy of Barnaby Norris.)

Figure 11. At the other end of the 40-metre fibre length, AAO astronomer Rob Sharp peers into one of the two CCD cameras of the AAOmega spectrograph in its enclosure deep in the dome. (Image courtesy of Barnaby Norris.)

bright to observe faint objects in visible light. Investigations as diverse as the characteristics of Venus' upper atmosphere, the weather on brown dwarf stars and filaments of galaxies in clusters – as well as survey-type work – are IRIS2's stock in trade.

The third string to the AAT's bow – also a bright-time instrument – is UCLES, the University College London Echelle Spectrograph. Like AAOmega, this substantial piece of kit occupies its own room on the AAT's observing floor, but starlight reaches it not by optical fibres but by a series of mirrors that steer the telescope beam into its enclosure. UCLES is a survey instrument too, but it is used to obtain very detailed spectra of stars, one at a time. It is perhaps most famous for its contribution to our knowledge of planets in orbit around other stars. The Anglo–Australian Planet Search programme has discovered some 10 per cent of all known extra-solar planets (currently around 330) by means of the 'doppler wobble' technique, which looks for stars being pulled to one side or the other by the gravitational attraction of their planets. UCLES has also been used for pioneering work in asteroseismology, in which minute oscillations in the surfaces of stars reveal details of their structure and age.

This core suite of high-tech instruments is occasionally supplemented by visitor instruments brought in by teams of observers from outside institutions. It's also highly dependent on the proper functioning of the telescope itself, and that requires regular maintenance and upgrades. A recent highlight was the replacement of the original telescope control computer (an Interdata 70, occupying four large electronics racks and still working perfectly after thirty-four years) by something the size of a laptop. The Interdata was of such significance – and in such good condition – that it now resides in Sydney's Powerhouse Museum.

That particular machine was very familiar to one of the early characters of the Anglo–Australian Observatory. Patrick Wallace, who later became Starlink Project Manager in the UK, wrote much of the software that still controls the AAT, but was equally famed for his practical jokes. These included sending a helium balloon painted with a face up to the prime-focus observing cage (Figure 12) where David Malin was working (history does not record what David shouted down the intercom), and the insertion of extremely rude words into some of David's famous star-trail photographs by means of a flashlight manipulated from the catwalk around the dome. Such stories are the stuff of legend in an organization like the AAO.

All the staff of the AAO, whether they be astronomers, instrument scientists, engineers, technicians (Figure 13) or administrators, contribute to the functioning of the observatory, and it is largely due to them that the institution has maintained a high record of productivity over the years. Repeatedly, in studies of the effectiveness of astronomical facilities worldwide, the AAT has come out at or near the top, and the most recent analysis of this kind has demonstrated that the strategies adopted a decade ago have paid off in keeping the telescope at the cutting edge of astronomy.

Published in 2008, this study examined the productivity (number of papers) and impact (number of citations – i.e. subsequent references in published works) of all major telescopes based on publications over the three years from 2001 to 2003. The analysis shows that the AAT is the first-ranked 4-metre telescope in the world, in both productivity and impact, achieving 2.3 times as many citations as its nearest competitor. Moreover, among optical telescopes of *any* size, on the ground or in space, the AAT is ranked fifth in productivity and impact.

As the AAO's Director, Matthew Colless, says, 'This is an

Figure 12. The prime-focus observing cage of the Anglo-Australian Telescope. (Image courtesy of Barnaby Norris.)

Figure 13. Dwarfed by the telescope structure, AAO technicians wind the armoured optical fibre cable onto its storage drums at the end of an AAOmega observing run. (Image courtesy of Barnaby Norris.)

extraordinary achievement.' It means the AAT's reputation is still jealously guarded, and the commitment and loyalty of today's staff suggest that it will continue, even as the 8-metre-class telescopes reach their maturity.

TOWARDS A NEW RÔLE

The Anglo–Australian Observatory has had five Directors throughout its history: Joe Wampler (1974–6), Don Morton (1976–86), Russell Cannon (1986–96), Brian Boyle (1996–2003) and, since 2004, Matthew Colless, who was formerly co-Principal Investigator on the 2dF Galaxy Redshift Survey. All have made their mark. Joe provided the initial momentum and brought the then technically advanced Image Dissector Scanner to the telescope. Don transformed the infant observatory into a mature and prestigious organization, while Russell stage-managed the absorption of the UK Schmidt Telescope into the AAO and initiated 2dF. Under Brian's leadership the AAO of the future was planned, with ever-more innovative use of the two telescopes and the signing of contracts to build leading-edge instrumentation for 8-metre-class telescopes. Matthew has continued this to good effect, but has also dealt elegantly and successfully with the wind-up of the Anglo–Australian Telescope Agreement. This is quite simply the biggest challenge in the observatory's history, representing a monumental change in funding, governance and administration. Matthew has also steered the AAO towards a much broader role.

When the Anglo–Australian Agreement comes to an end on 30 June 2010, the AAO will cease to be called the Anglo–Australian Observatory. The institution will certainly continue, but at the time of writing (February 2009), it's not clear what its new name will be. Most likely it will simply be 'AAO', and if you want to make it stand for something, it could be the Australian Astronomical Observatory, or the All-Australian Observatory, or even the Absolutely Australian Observatory. The main ingredient, of course, is the fact that the organization will be funded by the Australian government, rather than jointly with the British, as hitherto.

What has led to this change? Back in 2002 the United Kingdom became a partner in the European Southern Observatory, and the science agency for particle physics and astronomy at the time (PPARC) signified its wish to withdraw from the Anglo–Australian Telescope Agreement. The withdrawal has been staged, with a gradual reduction in UK funding being matched by a reduction in observing time. Because of the AAT's productivity, however, and the continuing high demand from its user community (there are still two-and-a-half times as many nights requested as are available), it is likely that British astronomers will continue to use the telescope for some time under special arrangements.

But the AAT is only part of the Anglo–Australian story. The UKST will remain part of AAO for as long as external funding allows it to be operated, either for the RAVE programme or in other future projects that require a telescope with a very wide field of view. The instrumentation group, once it has completed HERMES, will continue with contract work for other large telescopes, as it has done for the past decade. In particular, it is bidding to build WFMOS, a very substantial and innovative new instrument being considered for the Gemini or Subaru telescopes with the aim of understanding the mysterious dark energy that permeates the Universe, as well as further increasing our knowledge of the history of our own galaxy. Even more striking, perhaps, is a new potential rôle for the AAO as Australia's national optical observatory. Since January 2008, it has been the host organization for the Australian Gemini Office, which supports Australia's partnership in the two 8.1-metre Gemini telescopes in Hawaii and Chile. It is also involved in a formal collaboration with the two 6.5-metre Magellan Telescopes, located in Chile and operated by a consortium of US institutions. Thus, the model of the AAO as a centre for all of Australia's involvement in optical astronomy facilities already exists.

Through a collaboration involving the ANU and a consortium of research institutions called Astronomy Australia Ltd, Australia is a partner in the 24.5-metre (equivalent) Giant Magellan Telescope (GMT), which is currently in the design and development stage. As well as the prospect that AAO might help to construct large new auxiliary instruments for this telescope, there is the truly exciting possibility that the observatory could one day host the national office for GMT. The AAO's laboratory in Sydney would then become the focus for Australian involvement in the new generation of Extremely Large Telescopes. Clearly, there is much for the new breed of AAO astronomer to look forward to.

THE ALL-AUSTRALIAN ASTRONOMER

Despite the advent of web-based weather radar, satellite imaging and all-sky cameras, astronomers like to get a first-hand impression of the condition of the sky. Whether or not there is cloud, which way it is moving, the smell of rain on the wind, perhaps – signs that are impossible to read from a windowless control room (Figure 14), no matter how well equipped it is. The AAT is no exception, and its catwalk is a place that affords not just routine weather checks but some of the most magnificent views in the world.

By night, of course, the vault of the Warrumbungles' pollution-free sky is visible in all its breathtaking glory. Even when there is no moonlight, the stars of the Southern Hemisphere shine brightly enough to allow the handrail and flooring of the walkway to be plainly seen without a torch. Throughout much of the year, the Milky Way arcs from horizon to horizon in a gossamer band, and it's easy to visualize the flattened disk of our galaxy encircling the sky completely, if we could but see through the Earth's dark form. Science tells us that this glowing

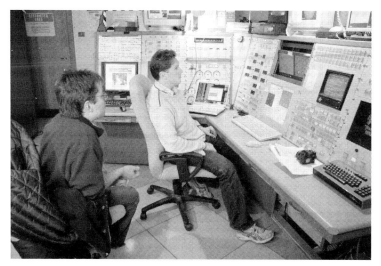

Figure 14. Where do you look through? AAO astronomer Rob Sharp (right) and Night Assistant/IT specialist Tim Connors contemplate AAOmega guide stars in the AAT control room. The last vestiges of the old Interdata 70 control system (now in a museum) are on the right. (Image courtesy of Barnaby Norris.)

swathe outlines the Solar System's home in the Universe. But from the catwalk of the AAT, it looks for all the world like the dreamtime river of Aboriginal legend, and it is easy to feel an affinity with the first humans who observed the sky from this place.

To some Aboriginal people, the Milky Way's dust clouds and dark lanes are the head, neck and body of an emu. To others, its two glowing companions represent an old man and an old woman sitting by a camp fire, which is the star we call Achernar. Nothing the unaided eye can see gives the remotest hint that these two Magellanic Clouds are whole galaxies of stars, albeit small ones, and that the light from the nearest of them has been on its way for 170,000 years. Aboriginal legend also has much to say about the stars themselves, as they dutifully mirror the rotation of the Earth in their nightly excursion around the sky. Stories of hunters, beautiful young women, sacred creatures and munificent spirits seem to complement science's view that they are just other suns, fellow travellers in the galaxy with our own. For perhaps 10,000 years, thinking people have watched the Universe from Siding Spring Mountain, and have been inspired by it. With an ancestry like that, the

All-Australian Astronomer will have much to be proud of, whatever organization he or she belongs to.

ACKNOWLEDGMENTS

I have always considered it a great privilege to work amongst the staff of the Anglo–Australian Observatory, and it is a pleasure to thank all present and former staff members for their contribution to the subject matter of this article, and for many conversations over the years. Special thanks go to the AAT Board and the AAO Director, Matthew Colless, for their support throughout, and to David Malin and Barnaby Norris for making their stunning images readily available. It's also a pleasure to acknowledge Roger Bell, former AAT Board Secretary, with whom I worked on a brief historical survey of the AAT for its twenty-fifth birthday. Finally, I thank Marnie Ogg for useful suggestions on the form of the article.

FURTHER READING

Frame, Tom and Faulkner, Don (2003), *Stromlo: An Australian Observatory*, Allen & Unwin, Sydney.

Gascoigne, S.C.B., Proust, K.M. and Robins, M.O. (1990), *The Creation of the Anglo–Australian Observatory*, Cambridge.

Watson, Fred (1995), 'The Universal Astronomer: David Allen, 1946–1994', in *1996 Yearbook of Astronomy* (ed. Patrick Moore), pp. 126–140, Macmillan, London.

Watson, Fred (2001), 'The Enduring Legacy of Bernhard Schmidt', in *2002 Yearbook of Astronomy* (ed. Sir Patrick Moore), pp. 224–42, Macmillan, London.

Watson, Fred (2004), *Stargazer: The Life and Times of the Telescope*, Allen & Unwin, Sydney.

The Apollo Lunar Surface Experiments Package

DAVID M. HARLAND

In 1961 President Kennedy set his nation the goal of landing a man on the Moon before the decade was out. Although he refrained from stating this to be a scientific venture, it was evident that such a mission would make a significant contribution to our understanding of the Moon.

ORIGIN

In 1964 several science planning teams comprising experts in a range of scientific disciplines compiled preliminary lists of instruments for Apollo astronauts to deploy on the Moon. On 7 June 1965 George E. Mueller, Director of the Office of Manned Space Flight at NASA headquarters, approved funding for the development of the Lunar Surface Experiments Package (LSEP). Since the temperature falls to $-200°C$ during the fortnight-long lunar night, it was decided that the package should use a radioisotope thermal generator in which thermocouples convert radiogenic heat into electricity. The package was to be capable of being offloaded from the lunar module and set up by two astronauts within an hour, and was to transmit data to Earth for at least one year. On 3 August NASA issued six-month study contracts of $500,000 to Bendix Systems, TRW Systems and the Space-General Corporation. In early 1966 the project was renamed the Apollo Lunar Surface Experiments Package (ALSEP). On 16 March 1966 NASA Administrator James E. Webb decided that, in view of the company's experience in supplying experiments for robotic lunar spacecraft, Bendix of Ann Arbor, Michigan, would receive the contract to design, manufacture, test and supply the ALSEP.

On 6 September 1968 Robert R. Gilruth, Director of the Manned

Spacecraft Center, argued against the ALSEP being assigned to the first lunar landing mission, which was to be essentially an engineering test flight. But on 9 October the Manned Space Flight Management Council at headquarters, chaired by Mueller, accepted a proposal by Wilmot N. Hess, Director of the Science and Applications Directorate of the Manned Spacecraft Center, to develop a smaller and somewhat simpler Early Apollo Scientific Experiments Package (EASEP) for this mission. For the purposes of this article, these latter instruments will be considered to be part of the ALSEP.

THE INSTRUMENT SUITE

The suite comprised a selection of modular instruments which could be mixed and matched to study the various themes of lunar science. Generally, later flights carried larger packages, and much of the first

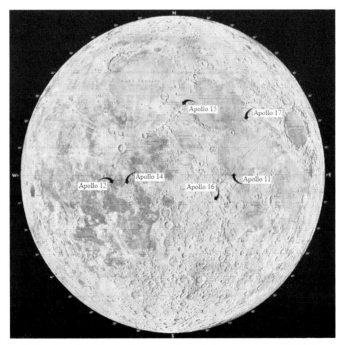

Figure 1. The sites where the Apollo missions landed. (Image courtesy of NASA.)

moonwalk was assigned to their deployment. Figure 1 shows the locations of the Apollo landing sites on the surface of the Moon. Some items were assigned to several missions in order to create 'networks' capable of coordinated measurements. Also, later flights were assigned instruments to follow up earlier results. Several independent themes were addressed in parallel. The main focus of investigation was the state of the Moon, but the space environment in its vicinity was also investigated and a number of other projects were pursued on an ad hoc basis.

The ALSEP suite

Apollo	11	12	13	14	15	16	17	Times
PSE	Y	Y	Y	Y	Y	Y	–	6
ASE	–	–	–	Y	–	Y	–	2
LSPE	–	–	–	–	–	–	Y	1
LTGE	–	–	–	–	–	–	Y	1
SEPE	–	–	–	–	–	–	Y	1
HFE	–	–	Y	–	Y	Y	Y	4
LNPE	–	–	–	–	–	–	Y	1
LSM	–	Y	–	–	Y	Y	–	3
LPM	–	–	–	Y	–	Y	–	2
CCGE	–	Y	Y	Y	Y	–	–	4
LACE	–	–	–	–	–	–	Y	1
LEAM	–	–	–	–	–	–	Y	1
SWS	–	Y	–	–	Y	–	–	2
SWCE	Y	Y	Y	Y	Y	Y	–	6
SIDE	–	Y	Y	Y	Y	–	–	4
CPLEE	–	–	Y	Y	–	–	–	2
CRDE	–	–	–	–	–	Y	–	1
LSCRE	–	–	–	–	–	–	Y	1
LRRR	Y	–	–	Y	Y	–	–	3
UVC	–	–	–	–	–	Y	–	1
LSG	–	–	–	–	–	–	Y	1

PASSIVE SEISMIC EXPERIMENT (PSE)

The solar-powered seismometer left by Apollo 11 (Figure 2) included a detector to measure the dust accumulation on and radiation damage to its solar cells, and an isotope heater to warm the electronics during the night. Despite operating temperatures that exceeded the planned maximum of +30°C, the instrument functioned normally throughout the period of maximum heating around local noon. Figure 3 shows the first signals from the Apollo 11 seismometer being received on Earth. With the output from the solar arrays in decline five hours prior to sunset, transmission was halted by command from Earth. It was reactivated early on the next lunar day, but the electronics had been damaged by the intense cold and the transmission was impaired. Near noon of its second lunar day the instrument ceased to accept commands, ending the experiment. Apollos 12, 14, 15 and 16 created

Figure 2. Buzz Aldrin stands beside the passive seismometer at the Apollo 11 landing site. The laser retro-reflector is behind the rod-antenna, both of which are aimed at Earth. (Image courtesy of NASA.)

Figure 3. Principal Investigator Gary V. Latham (left) watches as the first signal from the Apollo 11 seismometer starts to come in. (Image courtesy of NASA.)

a network of longer-term instruments to triangulate the sites of meteoroid strikes and moonquakes. By crashing spent vehicles on the Moon, it was possible to refine the calibration of the network; the discarded ascent stages of the lunar modules were crashed on to the Moon, starting with Apollo 12 (although not for Apollo 13 and in an uncontrolled manner on Apollo 16), and S-IVB rocket stages of the Saturn V launch vehicle were impacted, starting with Apollo 13.

In comparison to Earth, the Moon is almost seismically inert. Most of the events that were detected would be lost in the general 'noise' of the continuously adjusting terrestrial crust. The largest event was only magnitude 4 on the Richter scale. The largest impactor was calculated to be of the order of 5 tonnes in mass. It occurred on the far side, and the propagation of the seismic waves served to probe the internal structure of the Moon. The network showed that the Moon has a crust, a mantle and a core, and thus is a thermally differentiated body. The crust is plagioclase-rich and at an average of 70 kilometres thick is three times thicker than that of Earth. The interface between the crust and the mantle is sharp, as in the case of the Earth. The olivine and pyroxene mantle englobes a small core of roughly 25 per cent of the Moon's radius – in the case of Earth, the core is 50 per cent of the radius. The lunar crust is so deeply brecciated that it is very efficient at generating seismic reflections, with the result that energy is damped exceedingly slowly. As a result, the Moon 'rings' like a bell (Figure 4). Some internal events occur near the surface, but most originate at depths of 800 to 1,000 kilometres and are correlated with the tidal forces imposed on the Moon by its elliptical Earth orbit. The most active seismic epicentres are shown in Figure 5. The damping of seismic energy below a depth of 1,000 kilometres indicates that the rock is semi-molten, but this inner mantle is 'warm' rather than 'hot', and is too shallow for convection. Nevertheless, there may be isolated pockets of fluid in this zone, and the 'deep' seismic events may be due to movements of this magma in response to tidal forces. The internal structure of the Moon, as inferred from seismometry, is shown in Figure 6.

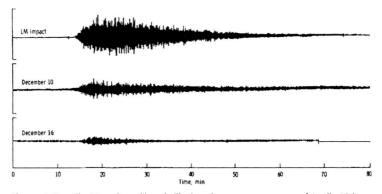

Figure 4. Top: The Moon 'rang like a bell' when the spent ascent stage of Apollo 12 lunar module was deliberately smashed on to the surface to provide a calibration signal for the seismometer. Below: The weaker signals from two later seismic events. (Image courtesy of NASA.)

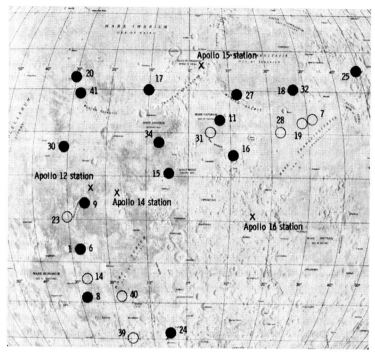

Figure 5. The most active seismic epicentres. A solid circle indicates a source whose depth was reliably determined – most of the order of 800 kilometres. The numbers are arbitrary identifiers. (Image courtesy of NASA.)

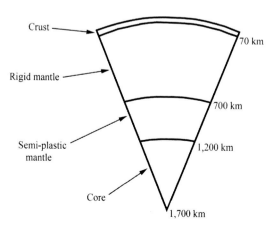

Figure 6. The interior structure of the Moon as inferred from seismometry. (By the author, based on data by NASA.)

ACTIVE SEISMIC EXPERIMENT (ASE)

The active seismic experiment probed the crust's upper kilometre at the landing site. It was operated by Apollos 14 and 16, both of which sampled ejecta blankets set in highland terrain. For one part of the experiment, once a line of geophones had been emplaced, an astronaut placed a 'thumper' down at specific points along the line and fired a

Figure 7. On Apollo 14, Ed Mitchell operates the 'thumper' along the string of sensors for the active seismic experiment. (Image courtesy of NASA.)

small charge (essentially a shotgun cartridge) to put a seismic signal into the ground (Figure 7). A mortar was also set up for use once the astronauts had returned to orbit, but Apollo 14's mortar was not fired and only three of the four charges of the Apollo 16 mortar were fired. The low speed of the seismic transmission (100–300 metres per second; less than in the Earth's crust) indicated the crustal material in both cases to be brecciated, with the Fra Mauro Formation at the Apollo 14 site being 75 metres thick and the Cayley Formation at the Apollo 16 site at least 100 metres in thickness. Both of these formations are believed to be ejecta from impacts by asteroidal-sized bodies early in the Moon's history.

LUNAR SEISMIC PROFILING EXPERIMENT (LSPE)

As a follow-up to the Active Seismic Experiment's mortar, the Apollo 17 astronauts installed geophones at the ALSEP site and then emplaced eight much larger charges at widely distributed points. When these charges were fired after the crew had left the Moon, the data revealed the floor of the mountain valley that they visited to be a slab of basalt at least 1 kilometre thick.

LUNAR TRAVERSE GRAVIMETER EXPERIMENT (LTGE)

The Apollo 17 astronauts measured the gravitational field at various points to further probe the substructure of their valley, and the results indicated that the lava flow that formed the valley floor was 2 kilometres in thickness.

SURFACE ELECTRICAL PROPERTIES EXPERIMENT (SEPE)

Apollo 17's Lunar Roving Vehicle was equipped to record radio signals broadcast by a transmitter that was laid on the ground at the landing site, to obtain information on the electrical conductivity of the subsurface to a depth of several kilometres. Its function was to calibrate an instrument carried by the mother ship probing the orbital ground track.

HEAT-FLOW EXPERIMENT (HFE)

One of the most important measurements that can be made of a planet is the rate at which it is losing heat to space. Once it had been established that the maria were lava extrusions, it was accepted that the Moon's interior is thermally differentiated. This experiment was to study the lunar 'heat engine'. It required strings of thermal sensors to be placed in holes drilled to depths of several metres. It was first assigned to Apollo 13, but this mission did not reach the Moon. On Apollo 15 a design flaw in the drill stems made it impracticable to reach the intended depth through a slab of consolidated material (Figure 8).

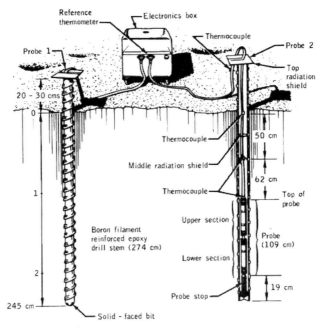

Figure 8. A diagram of the Heat-Flow Experiment as installed by Apollo 15. Owing to a flaw in the design of the drill stems, the probes could not be inserted as deep as planned. (Image courtesy of NASA.)

On Apollo 16 the experiment was lost when the cable was accidentally damaged. Fortunately, on Apollo 17 everything went well.

The lunar surface is a harsh thermal environment, being baked when the Sun is in the sky and frozen when it is not. The experiment showed that this month-long cycle affects only the upper half-metre of regolith. The existence of an isothermal layer at shallow depth had been inferred from near-infrared observations a decade earlier. Just below this, there is a build-up of heat leaking from the interior. At a depth of 1.5 metres the instrument showed the temperature to be a constant $-20°C$. The heat-flow rates of 21 mW/m^2 at the Apollo 15 site and of 16 mW/m^2 at the Apollo 17 site were surprisingly high (since the average for Earth is 87 mW/m^2), but because both sites are lava flows on the periphery of impact basins it is possible that the substructure imparted a bias – i.e. as much as 10 per cent of the observed heat-flow could be due to the fact that lava flows are efficient thermal conductors.

LUNAR NEUTRON PROBE EXPERIMENT (LNPE)

When the core samples from the early Apollo missions were examined, it was found that radioactive isotopes were created in the regolith by the absorption of cosmic-ray neutrons. Since taking the 'deep core' for the Heat-Flow Experiment left a hole, it was decided that Apollo 17 should insert a sensor into this hole to determine the flux and energy spectrum of neutrons at different depths in the regolith. This would serve to calibrate the technique by which the relative abundance of isotopes was used to infer the rate at which the fragmental débris layer (or regolith) was 'turned over' (or gardened) by the ongoing rain of micrometeoroids. The degree of mixing by an impact depends on the size of the projectile. Although larger impacts have a greater effect and excavate material from greater depths, smaller impacts are more common. The core samples revealed that while the uppermost centimetre is turned over every million years or so, it takes 1,000 times longer to turn over the top metre of material. In the intervals between significant impacts, the lunar surface is a remarkably inert place.

LUNAR SURFACE MAGNETOMETER (LSM)

Apollos 12, 15 and 16 deployed instruments to measure the Moon's magnetic field (Figure 9). These had three arms, each of which was to measure a specific Cartesian component of the local magnetic field. It was evident from the start that if the Moon possessed a dipole field, it would be extremely weak. The data were to be compared with that from Explorer 35, which had entered lunar orbit in 1967 to study the lunar environment. This database of the ambient fields enabled the Moon's own field to be isolated from the background fields associated with the solar wind and the presence of Earth (Figure 10).

The dynamo of electrical currents flowing in the Earth's metallic core generates a dipole field. Although the Moon has a small core, this appears not to have produced a magnetic field. However, long-term measurements of the field at the lunar surface, as the Moon entered and left the region of the Earth's magnetosphere that is blown 'down stream' by the solar wind, allowed the electrical conductivity of the material inside the Moon to be inferred – a process called electromagnetic sounding. Electrical conductivity depends on both temperature and chemical composition, and while the data did not determine the internal constitution, it served to constrain models derived from other data. In particular, it placed a limit on the size of an

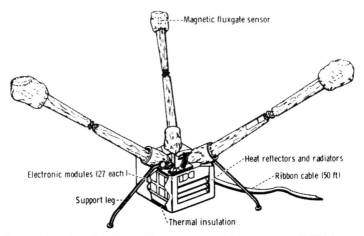

Figure 9. An outline of the lunar surface magnetometer. (Image courtesy of NASA.)

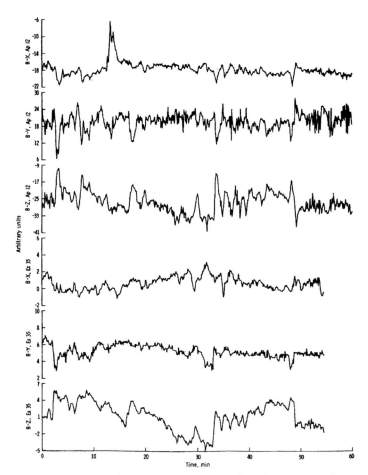

Figure 10. An example of 'squiggly line science' in the form of a comparison of data from the lunar surface magnetometer emplaced on the Moon by Apollo 12, and the data from Explorer 35, which flew in orbit around the Moon. (Image courtesy of NASA.)

iron core and on the temperature of the mantle. For an anorthositic crust over a mantle made primarily of the olivine and pyroxene (as suggested by surface sampling), the temperature would reach 1,000°C (sufficient to induce melting) at a depth in the range of 800 to 1,500 kilometres, with the actual depth depending upon the composition. This was consistent with the seismometry, which indicated a transition zone at a depth of 800 to 1,000 kilometres as the origin of deep moonquakes.

LUNAR PORTABLE MAGNETOMETER (LPM)

Following the discovery of a surprisingly strong magnetic field at the Apollo 12 site (Figure 11), it was decided to use a portable magneto-meter to take readings at widely separated points at a given site. In the case of Apollo 14, the field varied between 43 and 103 gammas between the landing site and the summit of the ridge 1 kilometre away, with the higher value being amongst boulders on the rim of a 370-metre-diameter crater. For Apollo 16 the field varied from 121 to 313 gammas. (To put this into context, the highest observed value is two orders of magnitude *lower* than at the Earth's surface.) On the local scale, the field at the lunar surface is derived from 'fossil' (or remanent) magnetism in the rocks, with the variation over short distances being due to whether the orientations of the individual rocks cause the fields to cancel out or to reinforce. The presence of individually magnetized rocks in a brecciated layer of impact debris suggested that

Figure 11. The ALSEP deployed by Apollo 12. The lunar surface magnetometer is in the foreground. (Image courtesy of NASA.)

when the Moon was young it generated a global dynamo field akin to that of Earth.

COLD-CATHODE GAUGE EXPERIMENT (CCGE)

This was set up by Apollos 12, 14 and 15 to measure the pressure of any gas at the lunar surface. Although it demonstrated that there was indeed an 'atmosphere', this was extremely tenuous (the pressure being fully fourteen orders of magnitude lower than at sea level on Earth) and the total amount of gas was no more than had been emitted by the lunar module in making its powered descent. The instrument was so sensitive that it could detect the presence of the astronauts by the water sublimated from their life-support backpacks.

LUNAR ATMOSPHERE COMPOSITION EXPERIMENT (LACE)

This mass spectrometer was set up by Apollo 17 to determine the composition of the gas at the lunar surface reported by the CCGE. It found the majority of the gas to be hydrogen, helium and neon. However, significant amounts of argon were detected at times of enhanced seismic activity. Since argon is the decay product of radioactive potassium, this could indicate ongoing venting of argon through fractures opened by the moonquakes. On the other hand, another instrument (SIDE) found a correlation with impact rates, implying that some of the gas was regolith material vaporized by meteoroids. Occasional detections of small amounts of methane, carbon dioxide, ammonia and water may have indicated the impact of small cometary fragments. Because the Moon's escape velocity is just 2.4 km/s, all but the heaviest gases will soon escape, and once they are ionized by solar ultraviolet they will be 'swept up' by the magnetic fields of the solar wind and carried away.

LUNAR EJECTA AND METEORITES EXPERIMENT (LEAM)

This experiment was set up by Apollo 17 to identify the nature of the small particles that strike the lunar surface. It employed a grid of plates to detect the trajectory and energy of impactors. It detected débris on almost horizontal low-energy trajectories from nearby impacts, but not the expected rain of cometary dust, nor any interstellar grains streaming through the Solar System.

SOLAR WIND SPECTROMETER (SWS)

This spectrometer was set up by Apollos 12 and 15 to record the energy density and temporal behaviour of the solar wind. Some 95 per cent of the solar wind is protons and free electrons, these being the products of ionizing hydrogen. The detector could determine the direction of the particles as well as their energies. At night, and when the Moon was in the 'tail' of the Earth's magnetosphere, the solar wind flux fell to zero.

SOLAR WIND COMPOSITION EXPERIMENT (SWCE)

Carried by all missions except Apollo 17, this metal-foil sheet was erected facing the Sun in order to trap solar wind ions. Because it was returned to Earth for analysis, the determination of the chemical composition of the solar wind was more accurate than could have been achieved by a detector that was left behind and reported its data by radio, but the exposure time was necessarily very short.

SUPRATHERMAL ION DETECTOR EXPERIMENT (SIDE)

This instrument was deployed by Apollos 12, 14 and 15 to detect ions with energies of less than 50 electron volts, corresponding to the *slowest* ions in the solar wind and gases at the lunar surface which had been ionized by solar ultraviolet. A correlation with the data from the seismometers suggested that at least some of the gas was regolith material that had been vaporized by impacts.

CHARGED PARTICLE LUNAR ENVIRONMENT EXPERIMENT (CPLEE)

Deployed by Apollo 14, this instrument supplemented SIDE by detecting ions in the 50–50,000 electron volts energy range.

COSMIC-RAY DETECTOR EXPERIMENT (CRDE) AND LUNAR SURFACE COSMIC-RAY EXPERIMENT (LSCRE)

Assigned to Apollos 16 and 17 respectively, these instruments detected ions in the energy range 100,000–150,000,000 electron volts, corresponding primarily to high-energy cosmic rays rather than the solar wind plasma. They took the form of 'plates' that were exposed on the Moon and returned to Earth for analysis of the 'tracks' left by the passage of relativistic ions.

LASER RANGING RETRO-REFLECTOR (LRRR)

A small laser reflector was left by Apollo 11, and larger units by Apollos 14 and 15 (Figure 12). The corner-cubed reflectors could return a laser to its source across a range of angles of incidence. While the signal strength was greatly diminished, because divergence in the laser beam in transit to the Moon meant that it illuminated a broad footprint, diluting the energy impinging on the instrument, the transit time of pulses measured the line-of-sight distance between the telescope firing the laser and the instrument on the Moon to unprecedented accuracy. Prior to the installation of retro-reflectors, the distance to the Moon could be determined only to an accuracy of 100 metres. In the 1970s it was possible to measure the distance to a given point to about 20 centimetres, but an accuracy of better than 1 centimetre can be achieved using today's technology.

By taking simultaneous measurements from a variety of sites it was

Figure 12. The Laser Ranging Retro-Reflector left on the Moon by Apollo 14. (Image courtesy of NASA.)

possible to determine the Moon's motion in three dimensions. Once the orbit was accurately known, it became possible to 'subtract' this in order to measure secondary effects, such as the tidal flexure of the lunar crust due to Earth's gravity during its perigee passage, and the diurnal flexure of the terrestrial surface as a result of lunar gravity. Long-term data established that the Moon is retreating from the Earth at 3.8 centimetres per year. The fine detail yielded information on the distribution of mass inside the Moon, indicating a core of 20 per cent of the lunar radius, which was consistent with the seismometry. Once the motion of the Moon was understood, it was possible to isolate terrestrial contributions to reveal variations in the Earth's rotation rate and the precession of its axis. Indeed, it became possible to directly measure the rates at which the continents are travelling as a result of plate tectonics.

FAR-UV CAMERA/SPECTROGRAPH (UVC)

Apollo 16 established the first astronomical observatory on the Moon by installing an ultraviolet camera (Figure 13). This comprised a small optical telescope and a spectrograph operating in the range of 500–1,600 ångstroms, a part of the electromagnetic spectrum that cannot penetrate the Earth's atmosphere. From a location in the lunar module's shadow, it automatically ran through a preprogrammed list of targets to investigate a variety of astronomical objects, the Earth and the lunar horizon.

Figure 13. The Far-UV Camera/Spectrograph in the shadow of Apollo 16's Lunar Module. (Image courtesy of NASA.)

LUNAR SURFACE GRAVIMETER (LSG)

If any experiment sent to the Moon had the potential to produce a Nobel Prize, it was this gravimeter. Although Einstein's theory of relativity predicted the existence of gravity waves, they had proved difficult to confirm. Working with a counterpart on Earth, any signal detected by just one instrument could not be a gravity wave, but any signal that was detected by both could *only* be a gravity wave passing through the Solar System. It was taken to the Moon by Apollo 17, but a flaw in the sensor's mechanism prevented it from attaining the required sensitivity. Nevertheless, it proved to be a capable seismometer and returned useful data of this nature.

MEN VERSUS ROBOTS

Some might wonder why astronauts were sent to deploy scientific instruments on the lunar surface. Might not it have been done more cheaply by automated landers? To answer this, one need only consider the difficulties that astronauts had to address in deploying their packages. Some instruments had to be set up in a specific orientation – for example, the solar panels of the passive seismometer that Buzz Aldrin set up required to be oriented to catch sunlight, which was fairly straightforward, and the instrument itself had to be precisely levelled, which proved difficult. The heat-flow experiments required to be emplaced in holes drilled to a depth of several metres. In the case of Apollo 15, Dave Scott found this inordinately difficult because of a flaw in the design of the drill stems. In contrast, on Apollo 16 Charlie Duke had no trouble in drilling, but the experiment was ruined when John Young snagged his boot on the cable and ripped it out of the electronics unit. And it was not just the instruments: on Apollo 12 the plutonium for the radioisotope thermal generator jammed as Al Bean attempted to remove it from its carriage flask, and was released only by Pete Conrad delivering a sharp blow to the flask using a hammer. Automated machines are good at carrying out specific tasks which go smoothly. It is difficult to design a machine capable of performing a multiplicity of specific tasks, and even harder to design one that can cope with problems. A human presence is the surest guarantee of success in such tasks. And, of course, sometimes a hammer helps too.

IN RETROSPECT

Although Kennedy called for the landing of a man on the Moon in the context of the Cold War, and critics of the Apollo programme dismiss it as an expensive exercise in 'footprints and flags', after the first landing had achieved the political objectives the rest of the programme was driven by scientific objectives. To members of the public, the most visible result of this work was the rocks that were collected by astronauts and returned to Earth for study. Many museums exhibit samples of lunar material. Unfortunately, the ALSEP results are less amenable to the public because they were generally squiggles on graphs. When funds for the technical and scientific support ran out on 30 September 1977, NASA shut down the ALSEP stations. Some of the instruments had suffered in the extreme environment, but useful data were still being provided. Including design and development, engineering support in Houston and the analysis of the data by laboratories around the world, the ALSEP cost around $200 million. As the operating budget was just $2 million per annum, in hindsight this penny-pinching was

Figure 14. The McDonald Laser Ranging Station (Image courtesy of University of Texas McDonald Observatory.)

shameful. In the ensuing years, scientists had due cause to rue this decision. Nevertheless, the laser reflectors, being inert, continue to facilitate geophysical research – notably involving the McDonald Observatory in Texas (Figure 14), the Apache Point Observatory in New Mexico and the Observatoire de la Côte D'Azur on the Plateau de Calern near Grasse in France.

FURTHER READING

McDonald Laser Ranging Station, University of Texas McDonald Observatory http://www.csr.utexas.edu/mlrs/

'ALSEP Termination Report' by J.R. Bates, W.W. Lauderdale and H. Kernaghan. Johnson Space Center, 1979 (Accession: 79N22979; Document ID: 19790014808; Report Number: NASA-RP-1036, S-480). This is available on the NASA Technical Report Server. To retrieve it, go to http://naca.larc.nasa.gov/search.jsp, enter 'aslep termination report' into the search box and download the resulting 165-page PDF, which is 200 megabytes.

Gamma-Ray Bursts

IAIN NICOLSON

Gamma-ray bursts are short-lived pulses of gamma radiation that occur at random locations in the sky and which are detected by satellites at an average rate of about one per day. The burst of gamma-ray emission

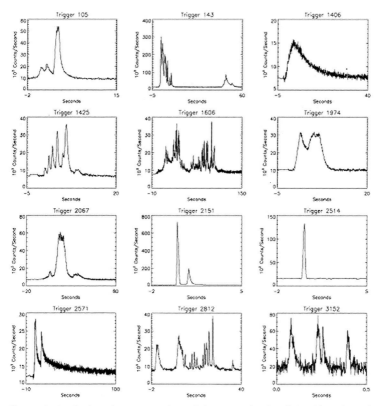

Figure 1. A selection of gamma-ray burst light curves showing their wide variety of durations and decay times. (Images courtesy of J.T. Bonnell (NASA/GSFC).)

can last from between a few milliseconds to a few hundred seconds (Figure 1), and is often followed by an 'afterglow' – a secondary outpouring of radiation that extends across the spectrum from X-rays through visible light to radio waves – which can persist for hours, days, weeks or even months. Bursts are named according to the year, month and day of discovery; for example, the burst that was detected on 13 September 2008 was designated GRB 080913. If more than one burst is detected on the same day, the bursts are distinguished by adding a letter of the alphabet; for example, the second burst detected on 19 March 2008 was labelled GRB 080319B.

Gamma-ray bursts are now known to be extremely powerful events which, at the peak of their brilliancy, can pour out as much energy in one second as the Sun will do in its entire 10-billion-year lifetime, and which take place in remote galaxies, typically billions of light years away. But for nearly three decades after the first GRBs were discovered, astronomers had no firm idea as to how far away or how powerful these enigmatic objects actually were.

DISCOVERY AND EARLY INVESTIGATIONS

Gamma-ray bursts were first detected purely serendipitously in the late 1960s by the Vela series of satellites which had been launched by the US military to look for the tell-tale flashes of gamma radiation that would betray clandestine nuclear tests in the atmosphere or in space in violation of the Nuclear Test Ban Treaty of 1963. The Velas did indeed detect a number of flashes of gamma radiation which appeared to come from random directions in space, but in no way resembled the gamma-ray signature of a nuclear blast. The first such burst was recorded in July 1967 by the Vela 4 satellite, but this discovery, together with subsequent detections by the Vela 5 and 6 satellites, was not made public until the data were declassified in 1973, at which time researchers from Los Alamos Laboratory published details of sixteen gamma-ray bursts that had been detected by Vela 5 and 6 between 1969 and 1972.

During the next couple of decades, more gamma-ray bursts (henceforth in this article abbreviated to GRBs) were detected by satellites and interplanetary spacecraft, but there was little real advance in astronomers' understanding of the nature of these enigmatic sources until the 1990s, when results began to pour in from NASA's Compton Gamma-Ray Observatory (CGRO). Launched in 1991, and originally designed for a two-year operating lifetime, CGRO continued to function for a full nine years before being de-orbited and destroyed in the atmosphere. It carried an instrument called the Burst and Transient Source Experiment (BATSE) which, between 1991 and 2000, detected more than 2,700 GRBs and, crucially, showed that they appear to be distributed isotropically – that is, they occur at random times in random directions, essentially uniformly distributed across the sky.

Prior to the BATSE results, astronomers had no clear notion of how far away, or how powerful, these enigmatic objects were. Some astronomers suggested that GRBs might be relatively local phenomena which occurred in the vicinity of the Solar System; others had suggested they might be located in the outer regions of our galaxy (the Milky Way Galaxy); others again had proposed that they might emanate from sources billions of light years away in the remotest recesses of the observable universe. The fact that they were distributed so uniformly across the whole sky and not, for example, concentrated in the plane of the Milky Way, pointed firmly to their being remote objects at 'cosmological' distances. If that were so, GRBs would have to be astoundingly energetic in order to be detectable at such enormous distances.

In addition to providing strong evidence that GRBs were located at very great distances, results from BATSE enabled astronomers to identify two major categories of burst – long duration and short duration. Bursts are designated as 'long' if they last for more than 2 seconds and 'short' if they last for less than 2 seconds; the average durations of long and short bursts are 20–30 seconds and 200–300 milliseconds respectively. Although short-duration GRBs are substantially less luminous than long-duration ones, their output is dominated by more energetic ('hard') gamma-ray photons, whereas long bursts predominantly consist of less energetic, longer-wavelength gamma rays; for that reason, short bursts are sometimes referred to as 'short/hard GRBs' and long bursts as 'long/soft' GRBs (Figure 2).

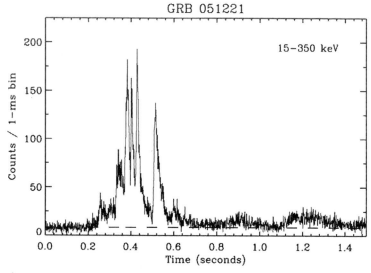

Figure 2. High resolution light curve of the short/hard burst GRB 051221A showing multiple peaks within the individual burst. (Image courtesy of NASA / Swift.)

SOLVING THE DISTANCE PROBLEM

In order to confirm that GRBs were indeed at 'cosmological' distances, and to gain a better understanding of their nature, astronomers needed to identify the sources within which they were embedded. The best hope for doing so hinged on trying to detect the longer-lived X-ray or optical afterglow which was expected to accompany the short-lived flash of gamma radiation. But BATSE was not itself equipped to make afterglow observations; nor did it (or any previous generations of gamma-ray detectors) have resolution which was good enough to pin down the positions of GRBs with sufficient precision to enable other instruments, in space or on the ground, to look for the afterglow or identify the sources within which the GRBs were embedded.

The vital breakthrough came in 1997, when the Dutch–Italian BeppoSAX satellite, which had been launched the previous year, finally succeeded in detecting the X-ray afterglow of a GRB and was able to identify the location of the burst with sufficient precision to enable ground-based telescopes and the Hubble Space Telescope to carry out follow-up observations at longer wavelengths. In addition to gamma-ray instrumentation, this satellite also carried an X-ray telescope which could promptly be turned to point to the location of any GRB detected by the on-board gamma-ray detector. When the gamma-ray detector registered a long-duration burst on 28 February 1997 (GRB 970228), its X-ray camera pinpointed an X-ray glow in the same location (Figure 3) and relayed its measured position to ground-based instruments and to the Hubble Space Telescope so that these instruments were able, for the first time, to image the anticipated afterglow, the peak energy of which shifted from X-ray to optical wavelengths and beyond as the afterglow aged and faded. Images obtained by ground-based optical telescopes revealed that the location of the afterglow coincided with a faint, fuzzy object which persisted after the afterglow itself had dimmed and vanished. HST images showed that this object looked like a distant galaxy (Figure 4), and that the GRB appeared to have flared up in the disc of the galaxy rather than at its centre, thereby ruling out the

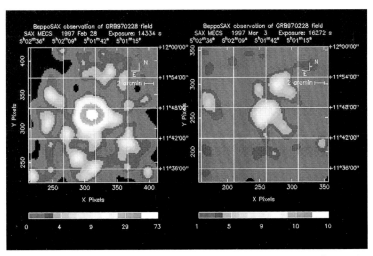

Figure 3. X-ray observations of GRB 970228 obtained by the BeppoSAX satellite. On the left is the original observation of the X-ray emission associated with the gamma-ray burst on 28 February 1997 and on the right is an image of the same region of sky on 3 March 1997, showing the extent to which the burst object had faded. The X-ray observations pinned down the position of the burst with sufficient accuracy to enable telescopes to detect the optical afterglow of a GRB for the first time. (Image courtesy of NASA / GSFC.)

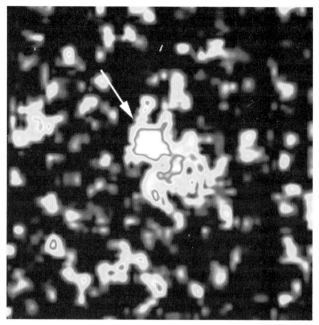

Figure 4. This Hubble Space Telescope image shows the visible fireball which accompanied the gamma ray burst GRB 970228. This was the first optical image ever taken which associated a gamma-ray burst source with a potential host galaxy at great distance. The arrow points to the fireball, which is a white blob immediately to the upper left of image centre. Immediately to the lower right of centre is an extended object (roughly resembling an 'E') interpreted to be the host galaxy where the gamma-ray burst is embedded. (Image courtesy of K. Sahu, M. Livio, L. Petro, D. Macchetto, STScI and NASA.)

possibility that this particular GRB might have been powered by a massive black hole in the centre of its host galaxy.

GRB970228 provided strong evidence in favour of the notion that GRBs were immensely powerful events that occur in distant galaxies. The clinching evidence came just three months later when BeppoSAX detected another long-duration burst – GRB 970508 – and its X-ray afterglow. Using very large ground-based telescopes (in particular, the 10-metre Keck telescopes in Hawaii), astronomers succeeded in measuring Doppler shifts of absorption and emission features in the spectrum of the optical afterglow which yielded a redshift of 0.835; this showed that GRB 970508 and its host galaxy lay at a distance of several billion light years, and implied that for a few seconds this burst had been a million times brighter than an entire galaxy. There could now be no room for doubt: gamma-ray bursts (or long-duration ones, at least) had been proved to be staggeringly luminous outbursts at cosmological distances.

CONTINUING INVESTIGATIONS

In the twelve years that have elapsed since that critical breakthrough, ongoing observations by a succession of spacecraft such as NASA's High-Energy Transient Explorer-2 (HETE-2) and the ESA's Integral óbservatory, gamma-ray detectors aboard interplanetary spacecraft and ground-based Imaging Atmospheric Čerenkov telescopes (which register flashes of light that are produced in Earth's atmosphere by incoming pulses of gamma radiation) plus follow-up observations from space and from a network of ground-based optical and radio telescopes, that have continued to detect more GRBs, have amply confirmed the immense power and remoteness of long-duration GRBs, and have broadened and deepened our knowledge and understanding of these extraordinary phenomena (Figure 5).

Currently leading the way on the observational front are the Swift satellite, which was launched in November 2004, and the Fermi gamma-ray space telescope (formerly known as the Gamma-ray Large Area Space Telescope – GLAST), which entered orbit in June 2008. Swift's Burst Alert Telescope (BAT), which has a wide field of view and is about five times as sensitive as BATSE, has been detecting about a hundred GRBs per year. As soon as the BAT detects a GRB it fixes its position to within four minutes of arc (approximately one-eighth of the apparent size of the Moon). The onboard X-ray telescope automatically swings to this position within about a minute and then determines the position of the burst to within about five seconds of arc. The third onboard instrument, the UV/Optical Telescope, then images the field of view and transmits a finder chart for the GRB to the ground so that a network of telescopes can home in on the afterglow (Figure 6). Key instruments aboard Fermi are the Burst Monitor, which detects around 200 GRBs per year, and the Large Area Telescope (LAT). The two instruments work in tandem so that as soon as the Burst Monitor picks up a GRB, the spacecraft slews round to allow the LAT to study it

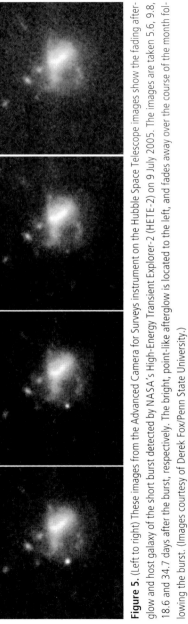

Figure 5. (Left to right) These images from the Advanced Camera for Surveys instrument on the Hubble Space Telescope images show the fading afterglow and host galaxy of the short burst detected by NASA's High-Energy Transient Explorer-2 (HETE-2) on 9 July 2005. The images are taken 5.6, 9.8, 18.6 and 34.7 days after the burst, respectively. The bright, point-like afterglow is located to the left, and fades away over the course of the month following the burst. (Images courtesy of Derek Fox/Penn State University.)

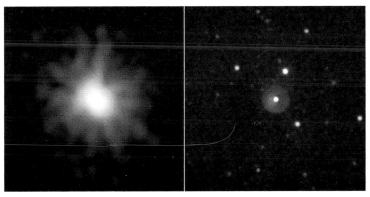

Figure 6. On 19 March 2008, Swift discovered a bright gamma-ray burst (GRB 080319B). The image shows the X-ray afterglow as seen by the X-Ray Telescope (left) and the bright optical afterglow as observed by the Ultraviolet/Optical Telescope on board the Swift satellite. (Image courtesy of NASA/Swift/Stefan Immler.)

in more detail. Fermi's burst monitor covers a much wider spread of energies, including higher energies, than Swift.

Among Swift's many and varied achievements have been the first detections, in 2005, of the afterglows of a short-duration GRBs, which generally are very much fainter than long-burst afterglows. Results obtained by HETE-2 and Swift have revealed that GRBs are more complex and diverse than can be described by a simple sub-division into long and short bursts, but have confirmed that short/hard bursts appear to be significantly different from long/soft bursts and almost certainly are triggered by a different mechanism from the one that powers long bursts.

Swift recorded the brightest burst ever seen (up to the time of writing) on 13 March 2008. The optical afterglow of this burst (GRB 080313B) was so bright that it briefly attained naked-eye visibility. At its peak, it reached an apparent magnitude of 5.3, significantly brighter than the usual naked-eye limit (in good dark skies) of magnitude 6; in principle, it could have been seen without optical aid for around 30–40 seconds had anyone been looking in the right direction at the right time. Spectroscopic measurements of the afterglow yielded a redshift of 0.937, which placed the GRB and its host galaxy at a distance of about 7.5 billion light years (Figure 7). About 3,000 times further away than the most distant objects that can normally be glimpsed with the naked

Figure 7. Peering across 7.5 billion light-years and halfway back to the Big Bang, NASA's Hubble Space Telescope imaged the fading optical counterpart of a powerful gamma ray burst, GRB 080319B, that holds the record for being the intrinsically brightest naked-eye object ever seen from Earth. For nearly a minute this single star was as bright as 10 million galaxies. Immediately after the explosion, the gamma-ray burst glowed as a dim 5th - magnitude 'star' in the constellation Boötes. (Image courtesy of NASA, ESA, N. Tanvir (University of Leicester), and A. Fruchter (STScI).)

eye under good conditions – the Andromeda Galaxy (M31) and the Triangulum spiral (M33) – GRB 080313B smashed the record for most-distant naked-eye object by a huge margin!

At the time of writing, Swift also holds the record for the most distant GRB ever measured. Detected on 13 September 2008, GRB 080913 has a measured redshift of 6.7, which corresponds to a distance of more than 12.8 billion light years. The event which generated this particular burst of gamma radiation took place about 800 million years after the Big Bang, and radiation from that event had travelled through space for more than 12.8 billion years before arriving at planet Earth.

THE SUPERNOVA LINK

The energies released by these events, if radiated equally in all directions, would be truly mind-boggling and exceedingly difficult to account for. For example, if the energy emitted by the most powerful long-duration GRBs had been radiated isotropically, the total output released in a few seconds would have been in excess of 10^{47} joules, which is about 1,000 times as much energy as the Sun will radiate in its entire lifetime. However, there is compelling evidence that indicates that the bulk of the energy emitted by a GRB is radiated in a relatively narrow conical beam, so that the total energy output is a fraction of what would be the case if the emission were isotropic (equal in all directions). Taking that into account, the total energy output, while still stupendous, is comparable with the output of an exceptionally powerful form of supernova – known as a hypernova – which is believed to occur when the core of an exceedingly massive star runs out of nuclear fuel and collapses to form a black hole rather than a neutron star (Figure 8). Hypernovae, which are thought to be about a hundred times as powerful as 'ordinary' supernovae, are believed to be exceedingly rare; perhaps only one supernova in 100,000 is as powerful as this.

The first observational evidence for a potential link between long-duration GRBs and supernovae was uncovered in 1998, when a highly luminous supernova, 1998bw, in a distant galaxy was detected remarkably close to the position of GRB 980425 and within about a day of the detection of that burst by BeppoSAX (the supernova lay within the 'error box' – the area of uncertainty associated with the measured position of the GRB).

Definitive proof of the link was provided by follow-up observations of the afterglow of GRB 030329, which had been discovered by the HETE-2 satellite. The optical spectrum and light curve turned out to be virtually identical to that of a supernova, and the X-ray spectrum showed the clear signature of oxygen atoms that had been excited and heated to very high temperatures by the blast wave of a supernova. GRB 030329 was exceptionally bright and its afterglow likewise

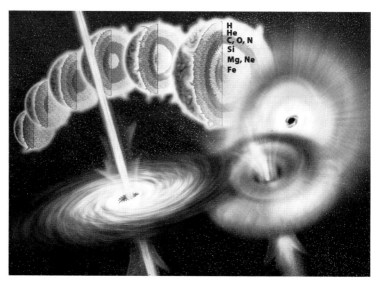

Figure 8. Stars shine by a process called nuclear fusion. This hydrogen 'burning' produces helium 'ash'. As the star runs out of hydrogen (and nears the end of its life), it begins burning helium. The ashes of helium burning, such as carbon and oxygen, also get burned. The end result of this fusion sequence is iron. Iron cannot be used for nuclear fuel. Without fuel, the star no longer has the energy to support the weight of its overlying layers. The core collapses. If the star is massive enough, the core will collapse into a black hole. The black hole quickly forms jets; and shock waves reverberating through the star ultimately blow apart the outer shells. Gamma-ray bursts are the beacons of star death and black hole birth. (Image courtesy of Nicolle Rager Fuller / NSF.)

because, at a distance of 'only' about 2.6 billion light years, it was one of the closest GRBs ever seen. Because of its brightness, astronomers were able to study its afterglow across a huge range of wavelengths in remarkable detail. In the weeks that followed the burst, spectral features due to an underlying supernova became clearer as the afterglow of the GRB faded. The underlying supernova, designated 2003dh, appeared closely similar to SN1998bw, a probable hypernova. GRB 030329 was the 'smoking gun' that established the connection between long GRBs and highly energetic supernovae.

POWERING THE BEAST

Currently, the front-running explanation for what powers long-duration GRBs is the collapsar/fireball model. According to this model, when the rapidly spinning core of an exceptionally massive star runs out of nuclear fuel, it collapses catastrophically – so quickly that the outer regions of the star are not immediately 'aware' of what has happened deep down in the core. The collapsing core creates a fast-spinning black hole (instead of the neutron star normally associated with a core collapse supernova) surrounded by an intensely hot and

Figure 9. A single frame from a computer animation of a gamma-ray burst destroying a star. This blue Wolf-Rayet star – containing about 10 solar masses' worth of helium, oxygen and heavier elements – has depleted its nuclear fuel. This has triggered a Type Ic supernova/ gamma-ray burst event. The core of the star has collapsed, without the star's outer part knowing. A black hole forms inside surrounded by a disk of accreting matter, and, within a few seconds, launches a jet of matter away from the black hole that ultimately made the gamma-ray burst. Here we see the jet (white plume) breaking through the outer shell of the star, about nine seconds after its creation. The jet of matter, in conjunction with vigorous winds of newly forged radioactive nickel-56 blowing off the disk inside, shatters the star within seconds. This shattering represents the supernova event. (Image courtesy of NASA/ SkyWorks Digital.)

rapidly spinning disc of gas (an accretion disc) composed of infalling stellar material. This whirling 'engine' channels material from the inner region of the accretion disc into two oppositely directed jets of matter and radiation that surge outwards at very large fractions of the speed of light perpendicular to the accretion disc and along the rotational axis of the central engine. These jets burst through the surface of the doomed star just before it is blown apart by an outward-moving blast wave (Figure 9).

The mix of electrons, positrons (antiparticles of electrons), protons and photons of which the jets are composed constitutes a 'fireball'. As faster-moving blobs within the jets overtake slower ones, the collisions create a series of shock waves that accelerate charged particles and generate gamma-ray photons within the expanding cone-shaped fireball. The gamma rays cannot escape until the temperature of the expanding fireball has dropped sufficiently for it to become transparent to these rays. The escaping burst of gamma rays then heads off in a narrow cone-shaped beam just ahead of the leading edge of the jet of matter, which rushes on outwards at a very large fraction of the speed of light (for example, the velocity of the central core of the jet associated with GRB 080313B is believed to have been about 99.99995 per cent of the velocity of light). Later on, as the jet ploughs into shells of gas previously ejected by the star, and into gas and dust in the surrounding interstellar medium, it creates the afterglow, radiation which declines in energy from X-ray through visible to radio, as the jet itself expands and slows down as a consequence of its interaction with the local environment.

An intriguing alternative scenario is the 'cannonball' (CB) model, whereby a series of blobs of magnetized, ionized matter (the cannonballs) are shot out from the central engine (a black hole or neutron star surrounded by an accretion disc, formed in an ordinary core-collapse supernova) along a common axis.

According to the cannonball model, a pulse of gamma rays is produced when the cannonball ploughs through a sea of photons which have been scattered by débris from the supernova or by previously ejected shells of matter. Photons colliding with electrons in the onrushing cannonball rebound with higher energies and so are converted into gamma rays (the process is called inverse Compton scattering) which are collimated into a narrow beam along the cannonball's direction of motion. The first phase of the afterglow consists of radiation emitted by the expanding, cooling cannonball; later on, most of the afterglow radiation is produced by electrons spiralling at large fractions of the speed of light within the cannonball's tangled magnetic field. Each pulse of emission within a GRB (bursts sometimes show several spikes) is generated by an individual cannonball.

Though proponents of the CB model contend that it requires fewer 'ad hoc' assumptions than the collapsar/fireball model, one observational argument that has been advanced against it is that high-resolution radio interferometry has found no evidence of the physical motion of the afterglow which ought to be detectable if the source were a cannonball moving off at a narrow angle to the line of sight. Both models have their pros and cons, and the real situation is undoubtedly more complex and subtle than models can currently describe, but the collapsar/fireball model is widely held to be the front-runner.

Either way, because the radiation is emitted in a narrow cone, a GRB will be seen only if that cone points towards the Earth. Consequently, we will be able to see only a relatively small fraction of all the GRBs that occur in the universe. If, as theory and observation suggest, the cone is only of the order of 10 degrees or so wide, we should expect to see only between 1 and 0.1 per cent of the total number of GRBs in the Universe.

SHORT-DURATION GRBS

Whereas the collapsar/fireball model works well for long-duration GRBs, a different kind of explanation appears to be needed to account for short GRBs. The total amount of energy radiated by short bursts is much less (by a factor in the region of 10 to 1000) than that emitted by long bursts, and their gamma-ray spectra are markedly different; the predominant energies of short-burst gamma-ray photons are higher than the predominant energies of the photons radiated by long-duration bursts. Short-burst afterglows, too, are much fainter, so much so that astronomers did not succeed in detecting them until 2005. The differences between the two types of burst were further highlighted when detailed studies of the fading afterglow of short-duration GRBs revealed no indication whatever of any associated supernovae; this argued strongly against the possibility that bursts of this kind are caused by the collapse of massive stars.

The favoured hypothesis is that the underlying engine that powers a short GRB is, once again, a black hole surrounded by an accretion disk. In this case it is formed not via the direct collapse of the core of a high-mass star but instead as the end result of the merger of a pair of closely orbiting neutron stars, or as the product of a black hole–neutron star merger (Figure 10). The end result would be a black hole surrounded by a rapidly spinning accretion disc composed of remnant debris from the merger – a system that would generate oppositely directed jets of matter hurtling outwards, perpendicular to the accretion disc, at a very large fraction of the speed of light. As with a long-burst GRB, the pulse of gamma rays would be radiated in a narrow cone by the jet, but, not as with a long-burst GRB, the jet would not have drilled its way out through the body of an exploding star.

Among the first few short GRBs for which afterglows were detected, two in particular (GRB 050709 and GRB 050724) showed distinctive features that appear to be consistent with black hole–neutron star mergers. Each displayed spikes of X-ray emission in the minutes and

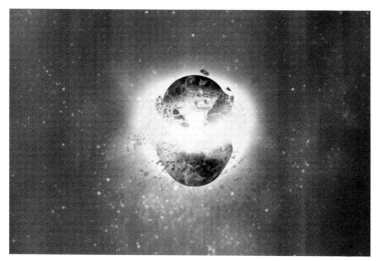

Figure 10. The most likely explanation for GRB 050709 is that it was produced by a collision of two neutron stars, or a neutron star and a black hole. Such a collision would result in the formation of a black hole (or a larger black hole), and could generate a beam of high-energy particles that could account for the powerful gamma-ray pulse as well as observed radio, optical and X-ray afterglows. (Image courtesy of NASA / D.Berry.)

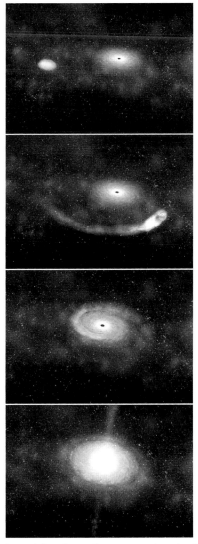

Figure 11. (Top to bottom) Four frames from a computer animation showing a black hole devouring a neutron star. Scientists say they have seen tantalizing, first-time evidence of a black hole eating a neutron star — first stretching the neutron star into a crescent, swallowing it, and then gulping up crumbs of the broken star in the minutes and hours that followed. (Images courtesy of Dana Berry / NASA.)

hours that followed initial gamma-ray burst – phenomena that could be explained by the black hole digesting residual blobs of material that had been left over when the black hole tore the infalling neutron star to shreds (Figure 11).

If indeed short bursts are triggered by mergers between closely orbiting pairs of neutron stars or black hole–neutron star pairs then, as the two compact objects spiral ever closer together and their orbital periods become shorter and shorter, they would be expected to radiate gravitational waves (disturbances in the gravitational field that propagate outwards at the speed of light) with a characteristic signature. Although nothing of this kind has been detected yet, facilities such as the Laser Interferometer Gravitational Wave Observatory (LIGO) or the proposed Laser Interferometer Space Array (LISA) may eventually be able to detect these tell-tale gravitational ripples. Where the burst is triggered by two compact objects spiralling together and merging, gravitational radiation should be detectable before the GRB takes place. With long-duration GRBs, the gravitational waves released when the collapsar creates a black hole ought to arrive at about the same time as the gamma-ray burst itself.

A THREAT TO LIFE?

In 1995, Steven Thorsett of Princeton University calculated that a nearby gamma-ray burst could wreak havoc with the biosphere if the Earth happened to lie in the direction of the beam of radiation emerging from its jet. Subsequent investigations by a team from the University of Kansas and NASA concluded that serious biological and climatic effect would be likely to occur if Earth were in the firing line of a GRB out to a range of about 10,000 light years, that such an event is quite likely to have occurred at least once within the past billion years and that certain distinctive features of the late Ordovician mass extinction, which took place about 440 million years ago, could be accounted for by an event of this kind.

Others have argued that because long GRBs occur predominantly in galaxies with youthful populations of stars and within which massive bouts of star formation are taking place, rather than in more 'sedate' and evolved galaxies such as our own, the odds on the Earth suffering a strike from the beam of a GRB are so small as to be not worth considering. By way of contrast, a more recent study by a team from the University of Sao Paulo, Brazil, has argued that a GRB could exert a lethal effect right across a galaxy out to distances of several hundred thousand light years.

Whatever the odds against such an event happening may be, being in the firing line of a nearby gamma-ray burst would be very bad news. The short gamma-ray burst itself would be unlikely to cause a major impact at ground level because practically all the gamma radiation would be absorbed in the atmosphere. However, that process would convert a significant fraction of the gamma rays into ultraviolet radiation, which would cause major damage to DNA and RNA at ground level. The burst of radiation would break apart atmospheric oxygen and nitrogen molecules and would initiate the formation of oxides of nitrogen, which would destroy a substantial fraction of the ozone layer, with consequent long-term damage to the biosphere and food chain.

COSMOLOGICAL PROBES

Because GRBs are so intensely luminous, they can be detected at huge distances and exceptionally high redshifts. They have already been detected out to redshifts as high as 6.7, which corresponds to looking back to an epoch just over 800 million years after the Big Bang; and there is every expectation that GRBs will soon be detected at even greater redshifts and distances. Because of this, there is intense interest in the possibility of using GRBs to test cosmological models, to investigate conditions in very early galaxies and to probe the distribution of matter in the intervening intergalactic medium.

Currently, the 'standard candles' (objects of known luminosity) which are most widely used to probe the distant universe and to test cosmological models are Type Ia supernovae (supernovae of this particular class attain closely similar peak luminosities). But the most distant Type Ia supernova yet discovered has a much more modest redshift of 1.7, and even the proposed Joint Dark Energy Mission spacecraft is unlikely to be able to detect Type Ia supernovae beyond redshift 2. Unfortunately, from the point of view of cosmologists, the energy outputs of GRBs differ from each other by a factor of up to 10,000, so that, on the face of it, they are anything but 'standard candles'. However, over the past few years astronomers have established correlations between the total energy outputs, or peak luminosities, of GRBs and a number of different observable quantities; for example, between the total energy of a burst and the way in which the brightness of the burst varies over time. With the aid of sets of correlations of this kind, astronomers have already been able to use GRBs, with some degree of success, to test cosmological theories and to place constraints on the relative proportions of matter and dark energy in the universe. While this technique cannot yet match the precision attained by more established cosmological probes, such as Type Ia supernovae and studies of the cosmic background radiation, its future looks promising.

Detailed observations of GRBs have already begun to yield intriguing insights into the properties of distant galaxies. A case in point is GRB 080607, which occurred in a star-forming galaxy some 11 billion light years away. Just twenty minutes after the burst had been detected by Swift, ground-based observers managed to obtain a spectrum of the afterglow using the giant 10-metre Keck Telescope. Imprinted on the afterglow were the clear spectral signatures of hydrogen and carbon monoxide, well-established constituents of the dense molecular clouds within which star formation takes place. The measurements showed that the burst must have occurred behind a dense, dust-laden molecular cloud that transmitted only about 1 per cent of the afterglow's light. Though many of the observed spectral lines remain to be identified, the results nevertheless reveal that this distant cloud contains proportions of chemical elements remarkably similar to what is observed in the Milky Way today. GRB 080607 has provided a unique insight into star formation at a time when the universe was about one-sixth of its present age, and has demonstrated in spectacular fashion the ability of GRBs to illuminate the properties of the distant universe in the remote past.

Gamma-ray bursts – arguably the biggest bangs since the Big Bang – are among the most dramatic and energetic phenomena in the Universe, fascinating in their own right and full of potential to probe the properties of the early universe and perhaps, eventually, to shed new light on the overall geometry, composition and evolution of the universe. Little wonder, then, that the study of GRBs remains one of the hottest research fields in contemporary astronomy (Figure 12).

Figure 12. This artist's impression shows the Fermi Gamma-ray Space Telescope in orbit, 550 kilometres above the Earth. The 4-tonne satellite, which was launched on 11 June 2008, was named in honour of the pioneering Italian-American high energy physicist, Enrico Fermi (1901–1954). (Image courtesy of NASA and General Dynamics.)

Hyperstar Imaging at the New Forest Observatory

GREG PARKER

Towards the beginning of March 2002 I purchased a beautiful Helios 6-inch refractor with a motorized mount in order to observe the heavens from my Brockenhurst garden – there was no New Forest Observatory with its 7-foot fibreglass dome back then. It was a conscious decision on my part to get a motorized – but not computerized – 'go to' mount, as I wanted to have the fun of finding the Messier and Caldwell objects all on my own, without any computer aid. Well, to cut a long story short, I didn't have a great deal of fun, meaning I didn't have a great deal of success in finding all those elusive objects, and I quickly realized I had made a very *big* mistake!

It's pointless to beat around the bush, so I decided I'd get one of those computerized Schmidt-Cassegrain telescopes to do my observing instead, and the choice for me was down to either a Celestron or a Meade. In fact for my finances at the time the choice was between either an 11-inch Celestron or a 12-inch Meade Schmidt-Cassegrain. The thing about the Celestron was that there were advertisements for this thing called (at the time) a Fastar. This was a very odd-looking beast indeed, a lens system that replaced the secondary mirror on the Celestron and turned the 'slow' native f/10 Schmidt-Cassegrain into an ultra-fast f/1.85 Schmidt-camera imager. Although I knew nothing about deep-sky imaging at the time, I felt that I really wanted the possibility of fitting this strange device at some point in the future, even if it was just to see how (and if) the thing worked. On this point alone I purchased the Celestron Nexstar 11 GPS scope during April 2002.

The night of Thursday, 2 May 2002, is one that is indelibly etched into my mind, and my soul. I had the Celestron Nexstar 11 GPS two-star aligned, in Alt-Az mode, with the GPS on, and I was ready to start observing all these objects I'd heard about but never seen before. Not knowing what would be visible that night, I simply typed 'M3' into the hand controller unit, and off the telescope went. The motors whirred for maybe a minute, the last little whirr being used to take up the gears' backlash. Expecting to see nothing, I peered through the eyepiece. Slap bang in the middle of the field of view was a wondrous glowing ball of stars, the first globular cluster I had ever seen. Never having seen this object or its like before, I thought it must have been a fluke. Pushing my luck further, I typed in another random number: this time I tried M88. Again, off went the telescope and this time a galaxy appeared at the centre of the eyepiece. Now although several decades of scientific training should have told me I couldn't have been so lucky twice in a row, I still wasn't entirely convinced, so I went indoors and brought out my copy of *Norton's Star Atlas*. I knew its time would come! Opening up the chapter on 'Interesting Objects, Maps 9 and 10, Clusters, Nebulae and Galaxies', I typed in the first object on the page, NGC3132, and set the telescope running. As in the old computer game, nothing happened. The hand controller indicated that the object was below the horizon. I then went to the second object on the list, NGC 3242, a planetary nebula in Hydra called the Ghost of Jupiter. This time the motors whirred into life and the stars slewed across the field of view. The motors wound down and I looked through the eyepiece as the gear backlash was being taken up. An amazing glowing object quickly locked itself into the centre of the eyepiece. Clearly this was the Ghost of Jupiter, and I logged my first planetary nebula that night.

I was now like a kid in a sweetshop: I had a whole mass of clusters, nebulae and galaxies all ready for the viewing. All I had to do was type in the Messier or NGC number. This was so good it seemed unreal. During the next hour I worked my way down the page and logged twenty-seven out of the thirty-four listed objects – it was incredible! Getting greedy, I next turned to a different page of *Norton*'s and started on a new list. It was at this point that I discovered M13, the great globular cluster in Hercules, for the very first time. You just don't forget first encounters like that!

Just after midnight the sky clouded over and I had to stop. I might well have perished outside from the cold if I had not been forced to give up the evening's viewing. Some six years on I can still say that was the most amazing two-and-a-half hours of my life (so far).

I really got into observing and spent every minute I could at the eyepiece on every clear night that came along. I even bought a pair of binoviewers – which meant doubling up on all my eyepieces, to enhance the viewing experience. However, once all the initial excitement had died down, I became disappointed. There were perhaps only half-a-dozen or so objects that looked *really* good through an eyepiece, and I'm a little ashamed to say that I started to get bored.

I had already bought the first version of the Hyperstar lens from Starizona (Arizona, USA) for the C11 in anticipation of imaging deep-sky objects. The Hyperstar lens is a times-one corrector that takes the place of the secondary mirror in a Schmidt-Cassegrain and turns a native f/10 instrument into an f/1.85 Schmidt-camera. The Hyperstar acts as a corrector by producing a nice flat wave front from the incoming curved wave front which would not be suitable for projecting on to a flat CCD imaging chip. It is a Schmidt-camera and it is not possible to fit an eyepiece to the Hyperstar lens; it is designed to accommodate a CCD imager, so you don't 'look through' Hyperstar-fitted telescopes – they are only for imaging.

In November 2004 I purchased a Starlight Xpress SXV-H9C 1.4-megapixel one-shot colour CCD camera, which was very well optically matched to the Hyperstar. I bought the CCD (after some long hard saving) in November, as Orion would be making its appearance, and this was the first area I wanted to start imaging in. The very first image I took (telescope still in Alt-Az mode) was of a random region of sky, and the exposure time was probably around ten seconds. But the end result was phenomenal. More stars appeared on the monitor after that short ten-second exposure than I had ever seen through the eyepiece at any time before. I was instantly hooked, and I am very ashamed to say that I have never used an eyepiece and looked through the C11 since that day!

No matter. I had finally arrived where, I guess, I had always wanted to be. I could image the heavens, in colour, and in some detail – it was quite unbelievable. I won't bore you with the details of how I discovered that you needed to go equatorial and get a guide scope in order to take really nice deep-sky images, but instead I will fast forward to a rather upsetting time in my imaging career. Let me put things into perspective first. It was generally agreed that the Hyperstar (the first version that I was using) was unable to take good deep-sky images. People on the internet forums went on about collimation and other things I knew nothing about, but I just kept on imaging and turning out quite nice deep-space photographs. Then one evening I wanted to take an image of M81 and M82 together in the same frame, and the way they were oriented at the time meant that they just didn't quite fit in. No problem: I just reached inside the dew shield and physically rotated the whole Hyperstar assembly within the corrector plate. Took a sub-exposure, and – complete disaster! All the star shapes had gone funny and the image quality was simply awful. What I hadn't realized up until that point was that I had – by sheer good luck – placed the Hyperstar into the C11 in a perfectly collimated position (yes, I should be doing the Lottery every week) and now, by rotating the Hyperstar assembly, I had instantly thrown away that perfectly collimated system. I felt physically sick for around a week until common sense slowly returned.

I had an existence proof! At least there *was* a position for perfect collimation of the Hyperstar to the C11. It might be extremely difficult to find this position again, but at least it was there, and if I could physically push the Hyperstar assembly around within the corrector plate (there is a 1 mm clearance around the edge of the secondary mirror cell assembly to allow this), then maybe I could get back to a fully collimated system. But how could I possibly do this? At this point I took a metal drill to my beloved C11 (you can see how desperate I must have been) and fitted four screw rods, which could be used to push the secondary cell plus Hyperstar around within the corrector plate clearance. You can see the outcome of this drill work in Figure 1. It looks, and is, a very Heath-Robinson affair – but it worked. After a few fraught hours I had brought the system back into some sort of reasonable collimation and I could resume imaging. However, I never again achieved the perfect collimation that I had chanced upon right at the beginning, and I always had some coma to contend with at the corners of the field of view.

It was around this time, during spring 2005, that I teamed up with an expert image processor, Noel Carboni from Florida, USA, and together we started to turn out images from the Hyperstar system that were the best seen at the time. A three-frame Hyperstar 1 image of the Horsehead region can be seen in Figure 2, and a four-frame Hyperstar 1 image of the Pleiades is shown in Figure 3. Altogether some fifty plus Hyperstar 1 images formed the basis of two exhibitions, a BBC television *Inside Out* programme, a Meridian News television programme, several short radio programmes and articles in the *Daily Mail* and *Daily Express*, as well as a number of other 'glossy' magazines. All was proceeding extremely well – except that I was growing restless once

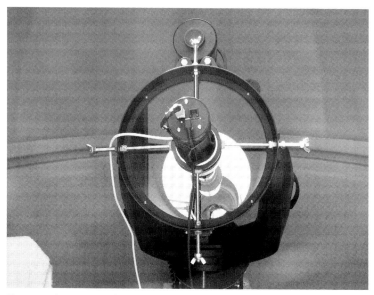

Figure 1. The outcome of the author's work when he took a metal drill to his C11 and fitted four screw rods which could be used to push the secondary cell plus Hyperstar around within the corrector plate clearance.

again. I wanted to try out the larger SXVF-M25C one-shot colour camera; this one has 6 megapixels instead of the 1.4 megapixels in the camera I had used up to this point. Also I was getting tired of the regular collimation the system needed, and the ever-present coma in the corners of the frame. At this point the Hyperstar and I parted company for just over a year.

In the *2009 Yearbook of Astronomy* you will see, in much more detail, a description of how I worked with the amazing Takahashi Sky 90 with an f/4.5 reducer/corrector and the SXVF-M25C camera. This almost perfect combination gave a huge (3.33 × 2.22 degrees) extremely flat field of view with perfect stars corner to corner; but it was very, very slow after the phenomenal speed of the f/1.85 Hyperstar. In fact, to be precise, it was a whole six times slower. Now, in order to get those high-quality deep-sky images I had to undertake imaging marathons. The two-frame RGB, H-alpha and OIII image of the Veil Nebula took over thirty hours of good-quality imaging time, and a similar amount of Noel's processing time. Although an almost perfect

Figure 2. A three-frame Hyperstar 1 image of the Horsehead Nebula region in Orion, produced when the author teamed up with expert image processor, Noel Carboni from Florida, U.S.A.

Figure 3. A four-frame Hyperstar 1 image of the Pleiades star cluster (M45) produced by the author and Noel Carboni.

imaging system, it was proving to be very expensive in terms of Noel's time as well as mine. In addition, my lovely C11 was now relegated to the role of a guide scope, which seemed more than a little perverse. Although we turned out some classic (in a couple of cases I would actually say definitive) images using the Sky 90 and M25C, including extremely deep versions of M31 (Figure 4) and M45 (Figure 5), I was slowly entering the imaging doldrums with this very 'slow' imaging system. I wouldn't bother to get started with an imaging session if I knew I didn't have at least four clear hours ahead of me, whereas with the old Hyperstar even an hour's worth of imaging between cloud banks was worthwhile for the high-quality data it could produce. I had no idea how I was going to resolve this huge problem.

In June 2008 I had to go to Arizona on a business trip for the University of Southampton – and Arizona, of course, means Starizona. To cut a long story short, Dean of Starizona shipped a new-version Hyperstar III (http://starizona.com/acb/hyperstar/index.aspx) and a MicroTouch auto-focusing system (http://starizona.com/acb/Micro

Figure 4. A deep version of the Andromeda Galaxy (M31), using the Sky 90 and M25C, produced by the author and Noel Carboni. The C11 was now relegated to the role of a guide scope!

Touch-Autofocusers-C745.aspx) to my hotel in Phoenix, and I returned to the UK once again with a Hyperstar. But this new Hyperstar III and my original version are chalk and cheese. The Hyperstar III readily accommodates the big SXVF-M25C chip, and it

Figure 5. A deep version of the Pleiades star cluster (M45), using the Sky 90 and M25C, produced by the author and Noel Carboni. However, the author was dissatisfied with this very 'slow' imaging system.

also comes along with collimation adjusters and a camera-angle adjuster as well. The field of view of the Hyperstar III with the M25C camera is a very respectable 2.4 × 1.6 degrees, with a sampling of 2.8 arc seconds per pixel. The MicroTouch auto-focusing system takes all the pain out of focusing the Hyperstar. I haven't mentioned up to this point that the critical depth of focus for the fast Hyperstar system is around 8 microns, equivalent to about one-tenth the diameter of a human hair! Trying to focus a Schmidt-Cassegrain system by winding the heavy primary mirror up and down a shaft, to an accuracy of within 8 microns, even with a manual electric focuser is a daunting task. It was not unusual for me to spend at least the first half-hour of an imaging session just trying to get a reasonable focus. The MicroTouch auto-focusing routine (computer-assisted, of course) takes all the pain out of focusing the Hyperstar, and within a couple of minutes it will zero in on a very precise focus. In fact, the focusing is now so quick and accurate that I am actually able to refocus during an imaging session. It seemed as if all the problems I had encountered with the original Hyperstar system had been addressed – but how well would the new system operate?

On 3 July 2008 I managed to obtain first light with the new Hyperstar III/C11 system. I chose the central region of NGC7000, the North America Nebula, for this initial outing. The result of just 56-minutes' worth of sub-exposures can be seen in Figure 6, and the result was truly astounding for such a short total imaging time. Collimation of the Hyperstar III system using the software package CCDInspector (http://www.ccdware.com/products/ccdinspector/) was also very straightforward and easy. In fact, in my very first attempt at collimating the Hyperstar I got a better result than I had ever achieved using the Sky 90 and M25C. The Hyperstar has returned to the New Forest Observatory and once again, together with Noel Carboni, we are turning out some ground-breaking images. For examples of some recent Hyperstar III work, please take a look at the Iris Nebula (Figure 7), the Cave Nebula region (Figure 8), the Coathanger Cluster (Figure 9) and the whole of the North America Nebula (Figure 10), which is a three-frame mosaic.

So is everything now settled in the New Forest Observatory equipment-wise, and have I reached the end of my journey in the

Figure 6. First light with the new Hyperstar III/C11 system. The central region of NGC7000, the North America Nebula, imaged by the author on 3 July 2008. This image – the result of just 56-minutes' worth of sub-exposures – was truly astounding for such a short total imaging time. Image processing by Noel Carboni.

Figure 7. Hyperstar III image of the Iris Nebula, produced by the author and Noel Carboni. Collimation of the Hyperstar III system using the software package CCDInspector was straightforward and easy.

Figure 8. Hyperstar III image of the Cave Nebula region, produced by the author and Noel Carboni.

Figure 9. Hyperstar III image of the Coathanger Cluster, produced by the author and Noel Carboni.

Figure 10. Hyperstar III image revealing the whole of the North America Nebula, which is a 3-frame mosaic produced by the author and Noel Carboni.

search for the perfect imaging system? Of course not! Filter changing requires breaking into the optical train, which in turn means recollimation each time a filter is replaced. A side-mounting filter system is required to circumvent this problem, and Dean at Starizona is currently working on this re-design. Then, of course, there are now bigger imagers than the M25C available. We can now get very nice 35 mm format imagers with 16-megapixels at our disposal. Could one of these cameras be Hyperstarred? Unfortunately that is not physically possible within the constraints of the C11's optical system – but then I guess, one can always be too greedy. Whatever the future holds regarding new developments (such as the mini-WASP array – http://www.newforestobservatory.com/index.php/category/writing/projects/mini-wasp-project/), I am more than happy with the performance of the current Hyperstar III system and will be keeping this one running at the New Forest Observatory for quite some time to come. You can visit the Observatory for all the latest news at http://www.newforest observatory.com/.

James Glaisher: Astronomer, Meteorologist and Aeronaut

ALLAN CHAPMAN

The nineteenth century saw the coming-together of many branches of science to lay the foundations of the world that we know today. Precision industrial engineering, high-grade optics, photography, electromagnetic registration, the infancy of photometry and the statistical analysis of mountains of new data all helped to transform and integrate the physical sciences. And in many respects, astronomy led the way.

So much of this innovation was pioneered by independent 'Grand Amateur' astronomers, though Great Britain had one outstanding public scientific institution which made major contributions to the advancement of astronomy and its related sciences: the Royal Observatory, Greenwich. In 1835, the running of the observatory was transformed when the dynamic, 34-year-old Cambridge professor George Biddell Airy (later Sir George), was appointed Astronomer Royal. And one gifted young man whom he brought down with him from the staff of the Cambridge Observatory to occupy what would become a very senior post at Greenwich was James Glaisher.

A NEW VISION FOR THE ROYAL OBSERVATORY

Airy realized that to fulfil its seventeenth-century Charter obligations the Royal Observatory had to make astronomical observations of ever-increasing accuracy, with a developing set of superbly engineered angle-measuring instruments, all of which, by 1850, he had designed personally. Long before this date the Royal Observatory had been routinely producing observations of such high accuracy as to easily fulfil its 1675 Foundation Charter requirement of enabling captains to find their longitude at sea. But Airy saw very clearly – as John Pond, his predecessor, had come to realize in the 1820s – that purely astronomical positions were not the only natural constants that one needed to measure and quantify if one were to adequately serve not only the Royal Navy and Merchant Marine but also the complex technical needs of the emerging industrial and imperial Britain. What about the Earth's complex magnetic field and the scientific study of the weather, both of which were seen as possessing some sort of connection to the Sun, Moon, and planets? It was hoped that the Royal Observatory could pin these connections down.

And what about the relationship between the weather and epidemic disease? In 1832 Britain, like many other European countries, had been struck by a new and alarming disease, Asiatic cholera. Prevailing medical opinion said that zymotic or contagious diseases were precipitated by environmental factors, especially the inter-relationship of air humidity, river and atmospheric temperature and barometric pressure. Many scientists and physicians, including Glaisher, were of the opinion that the nauseous blue mists that often hung above the heavily polluted River Thames had something to do with fever outbreaks. After the Register of Births, Marriages and Deaths was established by Act of Parliament in 1836, it came to be thought that the Royal Observatory should begin to keep a detailed record of these and other meteorological factors, as well as of the slight variations in the Earth's magnetic field. It was hoped that when all the data had been minutely collected, by means of precision instruments, and subjected to rigorous, mathematical analysis, clear physical laws would emerge, of a similar exactitude to those of Newtonian gravitation. Would we then be able to predict forthcoming epidemics with the same flawless accuracy as that used to predict the motions of astronomical bodies? And if we could predict when cholera or typhoid or scarlet fever were likely to strike from a given set of weather and environmental patterns, then we might be able to counter them more effectively.

Of course, we now know that Victorian physical and medical scientists were barking up the wrong tree, but it was not until 1854, and really around 1880, that John Snow, Louis Pasteur, Robert Koch and other medical scientists conclusively demonstrated that bacterial organisms, and in particular infected water supplies, not weather changes, caused zymotic diseases.

JAMES GLAISHER

Glaisher was born in Rotherhithe, in London's growing dock area, on 7 April 1809, the son of a watchmaker. The family later moved to Greenwich, where young James made the acquaintance of Mr William Richardson, one of the Royal Observatory Assistants under the then Astronomer Royal, John Pond, who was Airy's predecessor. And at the Observatory Glaisher became fascinated by precision astronomical instruments, and wanted to work with them. Although educated only in local schools, he clearly impressed the Greenwich astronomers sufficiently to be recommended at the age of twenty to Lt. Col. James to work on the Ordnance Survey of Ireland, of which James was Superintendent. James Glaisher's younger brother John had also been captivated by astronomy and precision instruments. Both young men devoted themselves to obtaining a thorough self-education, and by the time James went off to survey Ireland in 1829, his brother John was already working as a Computer, or junior mathematical assistant, at Greenwich. John Glaisher also worked at the Cambridge University Observatory, before becoming Astronomer at Dr John Lee's private observatory at Hartwell House, near Aylesbury. Sadly, he died of a kidney complaint in 1846.

James Glaisher was clearly fascinated by his survey work in Ireland, where he would have been involved with a team of men in dividing Ireland up into a series of exact interlocking triangles and meridian lines. Extremely precise theodolites, micrometers and measuring chains would have been used in this work, with distant cross-country sightings being made from the tops of church towers and mountains. Each datum, or trigonometrical point, would have to have been established astronomically with reference to observatories at Dunsink, Armagh, Greenwich and elsewhere, and the whole vast web of mathematical lines reduced to produce precise regional maps, and then a complete map of Ireland.

In this work, Glaisher would have not only mastered the essential practical skills of the cartographer and astronomer but also, one suspects, begun to acquire the administrative and organizational skills that were to play so conspicuous a part in his mature career. For in addition to being a scientist of outstanding ability, he was a natural leader and an efficient administrator.

Yet while James Glaisher seems to have possessed a constitution and a driving energy that would carry him into his ninety-fourth year, he had to withdraw from the Irish survey on health grounds. The prolonged exposure to the elements in that beautiful though often wet and windy land made him ill. One of his Irish experiences, however, was to have a formative influence upon the rest of his life's work and subsequent fame: his observation of the curious behaviour of clouds on Irish mountains, especially those of Galway and Limerick, from which, it was said, his interest in meteorology sprang.

In England, George Airy, who in 1833 was Plumian Professor of Astronomy at Cambridge and Director of the university's recently completed astronomical observatory, offered Glaisher a job as an Assistant Observer. Here, he truly cut his teeth as a working observatory astronomer, becoming accomplished in the taking of meridian transits in Right Ascension and Declination on the Cambridge Meridian Circle, and in using the equatorial refractor and micrometer to measure the coordinates of objects such as comets and perhaps minor planets (in which he had an abiding interest) and even binary stars. Binary star work was a cutting-edge branch of astronomical research in the 1830s, and no mathematical astronomer would have been unacquainted with it. Attention had first been drawn to close pairs of stars by Sir William Herschel in the 1780s, and in the intervening half-century many of Herschel's and subsequently discovered binary and triple systems had displayed movement around a baricentre. Now if these star positions could be measured with critical accuracy, perhaps to less than a single arc second, it would be possible to compute their respective masses, and – very significantly – to demonstrate that Newton's laws of gravitation operated in the remote stellar universe: a thing which was assumed, but could not be proved before the 1830s. Glaisher's Grand Amateur friends Dr John Lee and Samuel Whitbread were committed binary star astronomers, and he no doubt observed binary coordinates from their excellently equipped private observatories, at Aylesbury and at Cardington, Bedford, respectively.

At Cambridge Observatory, Glaisher distinguished himself by making a series of exact micrometric measurements through the equatorial of the position of Halley's Comet at its 1835 appearance. When Airy left Cambridge to become Astronomer Royal at Greenwich in 1835, he invited Glaisher to come with him, to an Assistantship at the Royal Observatory. Glaisher officially took up his post at

Greenwich in 1835, but his work on Halley's Comet at Cambridge was deemed so important that he was granted leave of absence to remain in Cambridge until February 1836, in order to complete it. He was to work at Greenwich for the next thirty-nine years, during which time he would rise from a trusted and capable Assistant Astronomer to a research scientist of truly international distinction, becoming a Fellow of the Royal Society by the time he was forty and subsequently Secretary and President of the Royal Astronomical Society (RAS), founder member of several new scientific societies and one of the 'great and good' of the British scientific 'establishment'.

GLAISHER AND AIRY

Glaisher's relationship with his boss, Sir George, was a complex one. Most of the men who made their careers at Greenwich between 1835 and 1881 (when Airy retired) were meticulous and clever, but they were rarely university graduates, and usually entered astronomy from local grammar schools at sixteen or seventeen years old, as mathematical 'apprentices'. Working hours could be long, especially if a man had a night observing shift, but qualified assistants got five or six weeks' paid leave a year, a salary of around £250 per annum, a pension, usually at sixty-five, and a very respectable position in society, roughly similar to that of a clergyman, local headmaster or bank manager. No one expected to make great discoveries: their job was rather to patiently measure the heavens to collect crucial data by means of which the Queen's ships could reliably find their way across any ocean on the face of the planet to within half a mile or less.

But Glaisher was an original, ambitious, daring lateral thinker who, as the years went by, developed a knack for using the Victorian 'mass media', by publishing books, academic research papers, and articles in magazines and newspapers, and by giving lectures. His opinions had a way of getting into the newspapers, and even while working as a full-time Greenwich Assistant, he was always being invited to sit on important committees – including one for the Great Exhibition of 1851. He was not, in many respects, an easy subordinate for Airy to work with, yet nonetheless the two men deeply respected and admired each other.

Although expecting loyalty, hard work and commitment from his staff, Airy never failed to do all he could to help them, and while, sadly, history came to think of Sir George Airy as a rather narrow bureaucratic Astronomer Royal, he possessed a remarkable generosity of spirit towards most members of his staff – for he himself had been a lad from modest origins, who had risen in the world on the back of a fortunate Cambridge scholarship. In addition to granting what must have been extra leave to Glaisher to fulfil his multifarious scientific engagements outside the Royal Observatory, Airy played a major part in Glaisher's election both to the Royal Society in 1849 (being his proposer) and to office in the Royal Astronomical Society, and – I suspect – in helping to secure for his son, James Whitbread Lee Glaisher, scholarships first to St Paul's Cathedral School, and then to Trinity College, Cambridge, which had been Airy's college, and at which Airy's son Wilfred had studied.

But the coexistence of two powerful and original scientific personalities within the close confines of the Royal Observatory also led to periods of conspicuous friction, most notably the trifling circumstance that led to Glaisher's precipitate resignation in 1874, to which we will return anon.

THE MAGNETIC AND METEOROLOGICAL DEPARTMENT

By 1800 there was a large literature dealing with the Earth's magnetic field, some of the most brilliant observational work on which had been accomplished by Airy's predecessor as Astronomer Royal, Edmond Halley, in the 1680s and 1690s. It was well known that the magnetic field acted laterally, drawing the compass needle to the north, while the magnetic 'dip' seemed to be a force that acted vertically and arched between the north and south poles. Furthermore, the field had long been known to have a westerly drift – though what scientists found baffling were the daily and longer-term small wobbles in the magnetic field. Could these be caused by astronomical factors, such as the Sun, or by solar or lunar gravitational cycles? There seemed to be only one way to find out: monitor them meticulously and then see if any 'laws' or astronomical correlations could be drawn from long runs of data. The great German scientists Alexander von Humboldt and Karl

Friedrich Gauss had suggested such a study across Europe, or even a global one, in the early nineteenth century.

Geomagnetism also took on a new urgency for the Royal Navy and Merchant Service in the 1830s. For while there were already ways of correcting the quarter-deck compasses of a great wooden man-o'-war to allow for the disturbing effect caused by a hundred great iron cannon, such as those carried on ships like HMS *Victory*, no one knew exactly what effects the new friction-generating steam engines and iron hulls of certain experimental vessels would have on the compasses. But before thoroughgoing shipboard experiments could be conducted, it was necessary to quantify the Earth's magnetic field from fixed observatories – in exactly the same way as one mapped the stars.

As we saw above, medical thinking by the mid-1830s was moving towards the establishment of a national meteorological observatory, to see what weather conditions were likely to cause an outbreak of cholera or scarlet fever. And where better could these new observatories be located (though there was already a small magnetic observatory at Kew) than at Greenwich, under the broad supervision of the Astronomer Royal? The authorization of and funding for what would become the 'Mag. & Met.' at Greenwich came in 1838, and the man whom Airy decided to appoint as Superintendent was James Glaisher. From then on, while he never ceased to be an astronomer and a leading Fellow of the RAS, Glaisher's career, especially after 1840, came to be, in the minds of both the scientific world and of the general public, inextricably connected with the weather.

By the early 1840s, new and specialist buildings were being put up within the grounds of the Royal Observatory, the most peculiar of which was a single-storey cross-shaped wooden structure in which to house the magnetic instruments. When Frederick Knight Hunt wrote an account of the Royal Observatory for Charles Dickens' popular family paper *Household Words*, in 1850, he told readers that no iron whatsoever, not even a single nail, was used in the construction, for even a single steel pin would have disturbed the ultra-precise magnets within. These buildings contained a set of delicate magnets suspended north–south, east–west and also in the vertical to measure the 'dip', on threads of untwisted silk, enclosed within darkened metal casings. Their movements were read by shining fixed narrow pencils of focused lamplight on to mirror spots on the magnets, and reading the place of the reflected light ray on a distant scale. These 'weight-less pointers' of light were capable of registering the slightest fluctuations in the positions of the magnets, caused by changes in the earth's magnetic field. In addition, a tall wooden mast carried a delicate electroscope – kept permanently warm and dry by a small lamp – which recorded changes in the ambient electricity of the atmosphere. (According to local East End folklore, the electroscope heating lamp was a signal light set up by the kindly Astronomer Royal to guide ships up the Thames at night – which says something about the dark skies of the 1850s. Just imagine trying to spot an oil-lamp flame on top of Greenwich Hill from a mile down river today!)

The permanent run of geomagnetic data gathered in the Magnetic Observatory could also be used to provide permanent magnetic marker points at the Admiralty's Compass Station, at Charlton, 2½ miles down river from Greenwich, where Her Majesty's ships were fitted with their compasses, 'swung' (or turned through 360 degrees) and given a magnetic correction at the start of a commission, which would, it was hoped, make their navigation safe wherever in the world they sailed. This was a perfect example, in fact, of how progress in scientific geo-magnetism was making its own contribution to rendering navigation safer and more exact.

Supervising and analysing the magnetic data was one of James Glaisher's duties after 1838, but the one that most obviously caught the public imagination was the Magnetic Department's companion institution, that of Meteorology.

Of course, keeping weather records was nothing new. The thermo-meter, barometer, hygrometer and rain and wind gauges all went back to the mid-seventeenth century, and many people thereafter across Europe had been keeping weather diaries. Indeed, Oxford University had kept a pretty much continuous weather record since 1767, and the new Armagh Observatory in Northern Ireland had maintained a detailed one since about 1817. What is more, each of the Astronomers Royal since Flamsteed in 1675 had kept weather records in varying levels of detail, and after James Bradley realized that barometric pressure, temperature and humidity affected the refractive index of the atmosphere, he had accompanied quadrant observations of a star in the meridian with a reading of his meteorological instruments. This meteorological data enabled Bradley to calculate how much distortion of position a star image might suffer at each single observation, until he worked out an air-refraction formula, which made it possible to establish a table of correction in order to quantify the air's distortion and obtain an exact angle.

But the significance of the new Meteorological Department at Greenwich after 1838 came from its sheer scale and thoroughness of operation. Not only were the instruments of the highest accuracy, but, as with the astronomical instruments, they could even cross-check one another: this was the case with, for instance aneroid and mercury barometers, and wet- and dry-bulb hygrometers of different design. Glaisher's assistants had to read each individual instrument several times each day, usually at two-hour intervals, day and night – as they did the magnetic instruments – to build up a massive body of data which could then be used for mathematical analysis, in order to extract the 'laws' that were believed to lie behind geomagnetism and meteorology, just as gravitation lay at the heart of physics.

With particular reference to cholera and other epidemics in London, Glaisher realized that the River Thames probably played a significant rôle, not so much because of the effluent that polluted it (though the 'morbid stench' of Old Father Thames, especially in summer, was empirically connected with disease), but because of its temperature changes. Apparatus was set up in and above the Thames on the Greenwich waterfront, including a moored vessel that could rise and fall with the tide, to record the respective temperatures of the river and the air above the river, to see if a set of causal factors could be found for cholera based on the heat and humidity of the river and its immediate environment.

All this regular monitoring of instruments, just like the measurement of Sun, Moon and star angles on the meridian, involved an enormous amount of sheer manual slog on the part of the junior assistants. Magnetic and meteorological observation would be revolutionized in the late 1840s, however, when Charles Brookes, a medical man with a passion for scientific instruments, invented a series of self-registering devices, using precision clockwork motors, 'weightless pointers' of light, and early photographic paper (photography was invented in 1838–9) to make the earth's magnetic field and the weather write what Charles Dickens called their own 'autobiographies'.

When Frederick Knight Hunt, the literary and medical friend of Charles Dickens, wrote his three articles on the Royal Observatory in *Household Words* in 1850, he had much to say about Mr Glaisher and the wonderful revelations that his new high-profile Royal Observatory department had brought forth. True, they had not found the cause of infectious disease, but improved public health and cleaner piped drinking water to the growing districts of London, with its population of almost 3 million, would soon be making their own impact in reducing cholera. Glaisher told Hunt and the readers of *Household Words* that existing statistical sequences indicated that when the air temperature in London fell from 45°F to 32°F, 300 additional people would die that week. And in what looks rather like a foretaste of global warming, he also pointed out that Greenwich Observatory weather records showed that, between 1771 and 1849, the average seasonal temperatures of southern England had risen perceptibly, which was considered to be a good thing, for a warmer climate was believed to provide a defence against disease.

In addition to his instrumental and statistical work at the Royal Observatory, Glaisher was all too well aware that a proper understanding of the weather required the collection of data, using instruments of standard accuracy and calibration, from across as wide a geographical area as possible. Towards this end, he tapped into Britain's great scientific resource: the dedicated amateurs. Amateur astronomers, naturalists, country vicars, doctors, schoolmasters and people in a variety of professions who loved practical science became part of Glaisher's volunteer army of meteorological data collectors, who eventually covered much of England, and later Ireland, Scotland and Wales. By 1870, ladies were also doing meteorological recording, including Miss Elizabeth Brown of Cirencester, Grand Amateur solar astronomer and meteorologist, who would be the driving force behind the founding of the British Astronomical Association in 1891.

Of course, these volunteers would not necessarily have done 24-hour monitoring, as at Greenwich, but they would usually read their instruments twice or so daily, and keep detailed logs. And once per week, they would fill in their Greenwich forms, and using the fast, efficient 'Penny Post', dispatch them overnight to the Royal Observatory. By 1855, moreover, not only was Britain becoming increasingly covered with an excellent railway and fast steam-packet system, making the central compilation of data physically easy, but the electric telegraphs that followed the railways and went under the sea to Ireland and France made it simple for a comfortably-off amateur scientist to send Glaisher a telegram if something big were happening, such as a freak storm hitting Ireland or Scotland, several hours before it reached the Royal Observatory. By 1860, the 'Victorian internet' of fast letter post and electric telegraphs was up and running, and Glaisher and Sir George Airy were among the first people to recognize its value for scientific purposes.

SEVEN MILES ABOVE WOLVERHAMPTON?

By the late 1850s, James Glaisher seems to have had a seat on virtually every committee that was involved in research into the physical sciences. And through the British Association for the Advancement of Science, with which both he and Airy were connected, he started to argue for the necessity of not just collecting meteorological data from stations across the Earth's surface, but investigating the upper atmosphere with a collection of scientific instruments. The association formed a Balloon Committee, and on 5 September 1862 James Glaisher and Henry Coxwell, the pilot of the new giant balloon of 90,000 cubic feet capacity (unnamed in 1862, but subsequently christened 'Mammoth'), set out on what was to become – even though there had been earlier flights – one of the most intrepid voyages into the unknown of the whole of the nineteenth century, when they cast off from Wolverhampton gasworks.

One may not think of Wolverhampton as the most obvious starting place for a great voyage of exploration, but it certainly was in 1862. Glaisher chose it as the place for their high-altitude balloon ascent for two reasons. Firstly, it was pretty well in the middle of Britain, so that the balloon was not likely to be blown out to sea. Secondly, Wolverhampton gasworks was capable of producing a particularly pure gas, which gave the maximum lift for each cubic foot of gas produced: aeronauts rarely used hydrogen by 1862, for coal gas, while not so good as hydrogen, was so much cheaper and much more easily available.

Glaisher and Coxwell, the aeronaut, took off from Wolverhampton gasworks at just after 1.00 p.m. on 5 September 1862. Fastened to a board in the large gondola, or basket, of the balloon were seventeen scientific instruments, which Glaisher monitored repeatedly as the balloon ascended, recording barometric pressure, temperature, humidity, sunlight and other factors as the air became thinner. Then, at 11,000 feet, the balloon broke through the cloud layer to 'surface' into brilliant sunshine, with a plateau of illuminated cloud below. Soon they had ascended 4 miles, and as they cast out more sand, they shot yet higher, rising 1 mile in ten minutes, and the sky overhead became a deep blue colour.

But as the atmosphere grew thinner, and they entered what we now know to be the jet stream, the pressure within the great bag of coal gas built up by the minute, increasing its lift and the danger of its splitting open, as the balloon continued to rise. No breathing apparatus existed in 1862, and Glaisher and Coxwell must have been fighting for their breath, yet still up they went, to learn whatever they could about the alien domain of intense cold, brilliant light and total silence into which they had ascended. Glaisher observed that his vision and co-ordination became disturbed. After 29,000 feet (5½ miles), and no longer able to read the instruments, he recorded that he slumped down helplessly into the balloon basket, and was unable to respond to Coxwell's attempts to revive him, though fully conscious and mentally alert. He knew that if the benumbed Coxwell could not let out a controlled amount of gas very soon, they would perish. Glaisher then became unconscious, but Coxwell, though numbed, frozen and dizzy, was still able to attempt to pull the valve cord to let out gas, to begin their descent. But the valve cord was stuck. Of course, Coxwell knew that if he could not release gas, then the balloon would ascend to a point where, as mentioned above, the pressurised gas bag would burst in the thin air, and they would fall like a stone. He therefore climbed up several feet into the rigging of the balloon, and because his hands were frozen and numb, took hold of the valve cord between his teeth, and pulled. Mercifully, the balloon began to descend. Glaisher came round, poured brandy upon Coxwell's and his own blackened hands to help restore the circulation – and no doubt drank some of the brandy as well – before immediately returning to his seventeen instruments and monitoring their behaviour during descent. Glaisher later calculated that he had been unconscious for some seven minutes, which gives some idea of the tremendous speed at which 'Mammoth' was ascending in what we now call the stratosphere.

The balloon made a safe landing at Cold Weston, near Clee St Margaret in Shropshire, at 2.50 p.m., after one of the shortest and most daring expeditions into the unknown ever undertaken by mankind. Far from being exhausted and needing to see a doctor, the 53-year-old Glaisher immediately set out on foot for Ludlow, over 7 miles away from the landing site, to hire a cart to convey the balloon to a railway station. Henry Coxwell stayed to guard the balloon and instruments, and started packing up while awaiting Glaisher's return.

The total 37,000-feet (7-mile) altitude was estimated using the rate of ascent and timed duration of the interval between Glaisher's last, 29,000-feet (5½-mile) reading and Coxwell's successful release of the valve ropes. Consequently, some scholars have been cautious in accepting the 37,000 feet figure, though few would deny that they exceeded an altitude of 6½ miles above Staffordshire and Shropshire. Coxwell, on the other hand, noted that when he was biting on the valve rope, the barometer registered only 7 inches of atmospheric pressure. That they continued to breathe, and suffered no permanent damage in such thin air, beggars belief! On some of his later flights, Glaisher did monitor medical and physiological responses to high altitude.

The 5 September 1862 flight was the most memorable of Glaisher's flights, but it was by no means his only one. Altogether, he made over twenty-eight flights. Some of these were high-altitude ones, though not approaching 37,000 feet; others were night flights to study the behaviour of the atmosphere after dark; and some were within moderate-altitude (4,000 feet or so) tethered balloons. All of them were for explicit scientific research purposes, and some were quite dangerous. On one flight, in September 1863, a storm suddenly sprang up, making for a very perilous landing. Although the balloon and all the scientific instruments on board were destroyed upon landing, Glaisher and the pilot escaped without serious injury. This truly was scientific exploration! Yet even when flying miles above the earth in a balloon, Glaisher's astronomical instincts were always at work; and in April 1863, he reported to the RAS his study of the 'Lines in the Solar Spectrum as observed in the Balloon Ascent of the 31st of March'. This wonderful combination of aeronautics, meteorology and the new science of solar physics probably makes Glaisher the first to do astronomical research from a 'flying machine'.

METEOROLOGY AND ASTRONOMY

Glaisher's 1862 ascent, not to mention the several others that he made before and after, confirmed his standing as a daring man of science and a public figure. They also taught us a lot about what was coming to be called 'the Ocean of the air'. But what did Glaisher's decades of meteorological and magnetic research (and not forgetting that of his colleagues in Great Britain and abroad) tell us about the great laws that lie at the heart of our terrestrial environment, and their relation to the wider gravitational forces which govern the Solar System?

By the early 1870s, the 'Great Meteorological Reduction' was in process, in which the now elderly Airy had set teams of human 'computers' the task of analysing decades of 'Mag. & Met.' data. The magnetic records appeared to indicate that solar activities and cycles had an influence upon the Earth's magnetic field. It also seemed that the atmosphere had tidal forces moving within it. Yet as far as being able to establish meteorological laws that were in any way comparable in exactitude to those of planetary physics was concerned, the Mag. & Met. work was deemed a failure.

In their passion for science, the Victorians may strike us as wildly optimistic. And while industrial manufacture, enormous advances in precision instrumentation and ingenious devices such as balloons, cameras and electrical circuitry had vastly increased mankind's capacity to scrutinize and record nature, they sometimes underestimated how very complex nature really is, and how elusive its underpinning laws can be. Yet without visionary projects like that of the Royal Observatory's Mag. & Met. Department, and without determined and exotic characters like James Glaisher, those foundations upon which the environmental sciences now stand would never have been laid.

We would never have worked out the nature of the intangible solar energy that bombards the Earth, causes the aurora borealis, sends ships' compasses haywire and disturbs telecommunications. In fact, it took scientists most of the twentieth century to make sense of that spectacular chaos system that we call 'weather'. But had early Victorian scientists in Britain and abroad not set about the monumental slog of recording every twist and twirl of our atmosphere, not to mention merrily risking their necks 7 miles above Wolverhampton, twenty-first-century science would have been much the poorer. And let us not forget that all of these pioneers were astronomers by original training, and were motivated by a desire to establish the relationship between celestial causes and terrestrial effects.

Yet while posterity has remembered Glaisher as a pioneer researcher in the sciences of geomagnetism, meteorology and atmospheric physics, one must bear in mind that he never ceased to be actively involved with astronomy. Minor planet astronomy continued to interest him: he travelled to Oundle, Northamptonshire, to observe the annular solar eclipse of April 1858, and communicated papers to the RAS on meteors and meteor showers.

JAMES GLAISHER AFTER GREENWICH, 1874–1903

Glaisher's sudden departure from his 39-year employment at the Royal Observatory was precipitated by what might strike us as an unbelievably trivial incident, though in fact it probably represents the 'straw that broke the camel's back' in Glaisher's often tense relationship with Sir George Airy.

On 5 September 1874 – twelve years to the day after his Wolverhampton balloon flight – the 65-year-old James Glaisher left work a few minutes early, prompting Sir George Airy to send him the following note of rebuke, which said: 'No other Officer of the Observatory leaves before 2 p.m.' This clearly touched a nerve in the senior and eminent Glaisher, who promptly responded to Airy on the same day, thus: 'Your fault-finding note to me this morning is so painful, that in consequence of it I will resign.' Glaisher then went on to remind Airy that he had almost completed his forty years' service, and that the difference between his computed pension owing and earned salary was less than £100 per annum – 'a sum I can earn out of the Observatory in a month'. Glaisher's official salary in 1874 was between £430 and £445, and if he was able to earn 'out of the Observatory' £100 *per month*, one can appreciate what an earning capacity he possessed. Glaisher would have made this income from writing and technical consultancy work and, one supposes, from picking up those paid invitations to submit an opinion which generally came the way of eminent people. This was the money that allowed him to live in a house on the fashionably elegant Dartmouth Terrace, Blackheath, and be a member of various dining and gentlemen's clubs in London. In fact, his total earned income from Greenwich and from his other activities was probably as large as, if not bigger than, the official salary of his boss, Sir George, who received around £1,100 per annum for being Astronomer Royal.

Yet it is clear that Airy meant no serious reprimand to Glaisher, for on the same afternoon, 5 September 1874, he sent Glaisher a most placatory and almost apologetic letter for having upset him, returning

Glaisher's resignation, pressing him to stay and concluding, 'I am, my dear Sir, Yours very truly, G.B. Airy.' But Glaisher had had enough, abided by his resignation and left the Royal Observatory on 31 December 1874. One senses that Sir George Airy's bark was worse than his bite, for it cannot be denied that the observatory held on to its staff for decades, with relatively few people choosing to leave. And while capable of being fastidious and bureaucratic, Airy was nonetheless a scrupulously fair-minded boss, who stood by his staff if they were loyal to him. But Glaisher had become too big a fish for the Greenwich Assistants' pool – and knew it. He had become a Great Man of Science in his own right.

Meteorology, geomagnetism and astronomy remained the principal foci of Glaisher's attention, but his range of interests was truly remarkable, as was the range of his friendships. For beyond the Greenwich Observatory, much of British astronomical research lay in the hands of the wealthy, independent Grand Amateurs, most of whom were very glad to know James Glaisher, FRS, and were keen to invite him to their homes and country-house scientific weekends. Of course, it was often these people who also did regional geomagnetic and meteorological work for the Greenwich Mag. & Met. Department, and one such place was Hartwell House, Aylesbury, the country seat of Dr John Lee, a wealthy London lawyer and Grand Amateur astronomer. The 'Albums' or visitors' books from Hartwell House are still preserved in Oxford, and Glaisher and his wife, Cecilia Belville Glaisher, often signed them when visiting. On 29 September 1856, they took along their eight-year-old prodigy son, future FRS and astronomer James Whitbread Lee Glaisher. Little James not only signed the Hartwell visitors' book but even included a letter to Dr Lee, written in Latin and signed 'Jacobus [James] W. Lee Glaisher'. Young James's other two Christian names derived from his father's friends Samuel Whitbread, the brewer astronomer of Cardington, Bedford, and Dr Lee.

James Glaisher, both before and after 1874, was a prodigious founder, president, secretary and treasurer of learned societies. He and a group of friends founded the Meteorological Society at Dr Lee's Aylesbury mansion and astronomical observatory in 1850, while Glaisher was a dominant figure in – and Charter President of in 1865 – the early Royal Microscopical Society, the Photographic Society and the Aeronautical Society. His interest in the history, topography and archaeology of the Holy Land also led him to play a significant rôle in the Palestine Exploration Fund, of which he was Chairman in 1880, and he became an authority on the meteorology of the region. In addition, he served the RAS both as Secretary and as President, in retirement from Greenwich.

James Glaisher is a fascinating figure in many ways. On the one hand, he was a classic example of a Victorian self-made man, rising from modest origins to an international standing in science. He was also an ingenious lateral thinker, in so far as he saw astronomy in the wider context of those gravitational and other forces that affected the Earth's climate and magnetic field, and went on to discover how those forces could be rendered of practical value, to navigation, and – potentially – to weather prediction. And in his fascination with astronomically induced natural cycles, including studies of geological Earth heat and solar heat, he was a major pioneer in the development of environmental and earth sciences. In terms of character and personality, he was a powerful, autocratic figure, and, as his daring high-altitude balloon ascents in the 1860s amply demonstrate, a man of action and great physical courage. Sadly, his relations with his wife Cecilia – herself a skilled microscopical draughtswoman and analyser of the structure of snowflakes – deteriorated, as Glaisher opposed their daughter's marriage, causing a serious breach within the family. Towards the end of his life, Glaisher, having left Dartmouth Terrace following Cecilia's death in 1892, lived partly at his new house and private meteorological observatory at Croydon, and partly with his son at Trinity College, Cambridge. It seems, however, that James and Cecilia Glaisher had been living apart for some years prior to 1892. James Glaisher's vigour remained into his early nineties, and he was still doing and publishing his meteorological researches at the age of ninety-three. He died in consequence of a stroke, on 7 February 1903.

ACKNOWLEDGEMENT

I am indebted to the sadly late Dr Mary Brück, who discovered and kindly informed me that the anonymous author of the *Household Words* articles about the Greenwich Observatory in 1850 was Frederick Knight Hunt.

FURTHER READING

Allan Chapman, *The Victorian Amateur Astronomer: Independent Astronomical Research in Britain, 1820–1920* (Praxis-Wiley, Chichester and New York, 1998).

Allan Chapman, 'The Observers Observed: Charles Dickens at the Royal Observatory, Greenwich', *The Antiquarian Astronomer* (Journal of the Society for the History of Astronomy), issue 2, December 2005, pp. 9–20.

The Edinburgh Review, April 1850, no. CLXXXIV: a group of six anonymous articles on the Royal Observatory, Greenwich, pp. 299–356.

'James Glaisher, 1809–1903', ed. H.P. Hollis; revision of earlier entry by J. Tucker, in *Oxford Dictionary of National Biography* (OUP, 2004).

'James Glaisher', obituary by 'W[illiam]. E[llis].', *Monthly Notices of the Royal Astronomical Society* LXIV (London, 1904), pp. 280–287.

Derek Howse, *Greenwich Observatory, vol. 3, The Buildings and Instruments* (Taylor and Francis, London, 1975).

Frederick Knight Hunt, 'The Planet Watchers of Greenwich', *Household Words* (ed. Charles Dickens), vol. I, no. 9, May 1850, pp. 200–204.

F.K. Hunt, 'Greenwich Weather Wisdom', *Household Words*, vol. I, no. 10, 1 June 1850, pp. 222–225.

F.K. Hunt, 'Swinging the Ship: A Visit to the Compass Observatory', *Household Words*, vol. I, no. 18, 27 July 1850, pp. 414–418.

John L. Hunt, OBE, 'James Glaisher, F.R.S. (1809–1903): Astronomer, Meteorologist, and a Pioneer of Weather Forecasting: "A Venturesome Victorian"', *Quarterly Journal of the Royal Astronomical Society* 37 (1996), pp. 315–347.

John L. Hunt, OBE, 'J.W.L. Glaisher, F.R.S., Sc.D. (1848–1928)', *Quarterly Journal of the Royal Astronomical Society* 37 (1996), pp. 743–757.

A.J. Meadows, *Greenwich Observatory, vol. 2, Recent History, 1836–1975* (Taylor and Francis, London, 1975).

L.T.C. Rolt, *The Aeronauts: A History of Ballooning, 1783–1903* (Longman, London, 1966).

Manuscript sources:

For Dr John Lee's 'Hartwell House Album' and J.W.L. Glaisher: Museum of the History of Science, Oxford, MS Gunther 10, 76–7 (Chapman, *Victorian Amateur Astronomer*, 86, 335, 338).

For Glaisher's resignation controversy with Airy: Royal Greenwich Observatory 6 (RGO6) archive, now deposited in Cambridge University Library, RGO6 MS 7/240–269 for sequence of letters, 5 Sept. – 31 Dec. 1874.

Part III

Miscellaneous

Some Interesting Variable Stars

JOHN ISLES

All variable stars are of potential interest, and hundreds of them can be observed with the slightest optical aid – even with a pair of binoculars. The stars in the list that follows include many that are popular with amateur observers, as well as some less-well-known objects that are nevertheless suitable for study visually. The periods and ranges of many variables are not constant from one cycle to another, and some are completely irregular.

Finder charts are given after the list for those stars marked with an asterisk. These charts are adapted with permission from those issued by the Variable Star Section of the British Astronomical Association. Apart from the eclipsing variables and others in which the light changes are purely a geometrical effect, variable stars can be divided broadly into two classes: the pulsating stars, and the eruptive or cataclysmic variables.

Mira (Omicron Ceti) is the best-known member of the long-period subclass of pulsating red-giant stars. The chart is suitable for use in estimating the magnitude of Mira when it reaches naked-eye brightness – typically from about a month before the predicted date of maximum until two or three months after maximum. Predictions for Mira and other stars of its class follow the section of finder charts.

The semi-regular variables are less predictable, and generally have smaller ranges. V Canum Venaticorum is one of the more reliable ones, with steady oscillations in a six-month cycle. Z Ursae Majoris, easily found with binoculars near Delta, has a large range, and often shows double maxima because of the presence of multiple periodicities in its light changes. The chart for Z is also suitable for observing another semi-regular star, RY Ursae Majoris. These semi-regular stars are mostly red giants or supergiants.

The RV Tauri stars are of earlier spectral class than the semi-regulars, and in a full cycle of variation they often show deep minima and double maxima that are separated by a secondary minimum. U Monocerotis is one of the brightest RV Tauri stars.

Among eruptive variable stars is the carbon-rich supergiant R Coronae Borealis. Its unpredictable eruptions cause it not to brighten but to fade. This happens when one of the sooty clouds that the star throws out from time to time happens to come in our direction and blots out most of the star's light from our view. Much of the time R Coronae is bright enough to be seen in binoculars, and the chart can be used to estimate its magnitude. During the deepest minima, however, the star needs a telescope of 25-cm or larger aperture to be detected.

CH Cygni is a symbiotic star – that is, a close binary comprising a red giant and a hot dwarf star that interact physically, giving rise to outbursts. The system also shows semi-regular oscillations, and sudden fades and rises that may be connected with eclipses.

Observers can follow the changes of these variable stars by using the comparison stars whose magnitudes are given below each chart. Observations of variable stars by amateurs are of scientific value, provided they are collected and made available for analysis. This is done by several organizations, including the British Astronomical Association (see the list of astronomical societies on p. 358), the American Association of Variable Star Observers (25 Birch Street, Cambridge, Massachusetts 02138), and the Royal Astronomical Society of New Zealand (PO Box 3181, Wellington).

Star	RA		Declination		Range	Type	Period	Spectrum
	h	m	°	′			(days)	
R Andromedae	00	24.0	+38	35	5.8−14.9	Mira	409	S
W Andromedae	02	17.6	+44	18	6.7−14.6	Mira	396	S
U Antliae	10	35.2	−39	34	5−6	Irregular	—	C
Theta Apodis	14	05.3	−76	48	5−7	Semi-regular	119	M
R Aquarii	23	43.8	−15	17	5.8−12.4	Symbiotic	387	M+Pec
T Aquarii	20	49.9	−05	09	7.2−14.2	Mira	202	M
R Aquilae	19	06.4	+08	14	5.5−12.0	Mira	284	M
V Aquilae	19	04.4	−05	41	6.6−8.4	Semi-regular	353	C
Eta Aquilae	19	52.5	+01	00	3.5−4.4	Cepheid	7.2	F−G
U Arae	17	53.6	−51	41	7.7−14.1	Mira	225	M
R Arietis	02	16.1	+25	03	7.4−13.7	Mira	187	M
U Arietis	03	11.0	+14	48	7.2−15.2	Mira	371	M
R Aurigae	05	17.3	+53	35	6.7−13.9	Mira	458	M
Epsilon Aurigae	05	02.0	+43	49	2.9−3.8	Algol	9892	F+B
R Boötis	14	37.2	+26	44	6.2−13.1	Mira	223	M

Star	RA		Declination		Range	Type	Period	Spectrum
	h	m	°	′			(days)	
X Camelopardalis	04	45.7	+75	06	7.4−14.2	Mira	144	K−M
R Cancri	08	16.6	+11	44	6.1−11.8	Mira	362	M
X Cancri	08	55.4	+17	14	5.6−7.5	Semi-regular	195?	C
R Canis Majoris	07	19.5	−16	24	5.7−6.3	Algol	1.1	F
VY Canis Majoris	07	23.0	−25	46	6.5−9.6	Unique	—	M
S Canis Minoris	07	32.7	+08	19	6.6−13.2	Mira	333	M
R Canum Ven.	13	49.0	+39	33	6.5−12.9	Mira	329	M
*V Canum Ven.	13	19.5	+45	32	6.5−8.6	Semi-regular	192	M
R Carinae	09	32.2	−62	47	3.9−10.5	Mira	309	M
S Carinae	10	09.4	−61	33	4.5−9.9	Mira	149	K−M
l Carinae	09	45.2	−62	30	3.3−4.2	Cepheid	35.5	F−K
Eta Carinae	10	45.1	−59	41	−0.8−7.9	Irregular	—	Pec
R Cassiopeiae	23	58.4	+51	24	4.7−13.5	Mira	430	M
S Cassiopeiae	01	19.7	+72	37	7.9−16.1	Mira	612	S
W Cassiopeiae	00	54.9	+58	34	7.8−12.5	Mira	406	C
Gamma Cas.	00	56.7	+60	43	1.6−3.0	Gamma Cas.	—	B
Rho Cassiopeiae	23	54.4	+57	30	4.1−6.2	Semi-regular	—	F−K
R Centauri	14	16.6	−59	55	5.3−11.8	Mira	546	M
S Centauri	12	24.6	−49	26	7−8	Semi-regular	65	C
T Centauri	13	41.8	−33	36	5.5−9.0	Semi-regular	90	K−M
S Cephei	21	35.2	+78	37	7.4−12.9	Mira	487	C
T Cephei	21	09.5	+68	29	5.2−11.3	Mira	388	M
Delta Cephei	22	29.2	+58	25	3.5−4.4	Cepheid	5.4	F−G
Mu Cephei	21	43.5	+58	47	3.4−5.1	Semi-regular	730	M
U Ceti	02	33.7	−13	09	6.8−13.4	Mira	235	M
W Ceti	00	02.1	−14	41	7.1−14.8	Mira	351	S
*Omicron Ceti	02	19.3	−02	59	2.0−10.1	Mira	332	M
R Chamaeleontis	08	21.8	−76	21	7.5−14.2	Mira	335	M
T Columbae	05	19.3	−33	42	6.6−12.7	Mira	226	M
R Comae Ber.	12	04.3	+18	47	7.1−14.6	Mira	363	M
*R Coronae Bor.	15	48.6	+28	09	5.7−14.8	R Coronae Bor.	—	C
S Coronae Bor.	15	21.4	+31	22	5.8−14.1	Mira	360	M
T Coronae Bor.	15	59.6	+25	55	2.0−10.8	Recurrent nova	—	M+Pec
V Coronae Bor.	15	49.5	+39	34	6.9−12.6	Mira	358	C
W Coronae Bor.	16	15.4	+37	48	7.8−14.3	Mira	238	M
R Corvi	12	19.6	−19	15	6.7−14.4	Mira	317	M
R Crucis	12	23.6	−61	38	6.4−7.2	Cepheid	5.8	F−G
R Cygni	19	36.8	+50	12	6.1−14.4	Mira	426	S
U Cygni	20	19.6	+47	54	5.9−12.1	Mira	463	C
W Cygni	21	36.0	+45	22	5.0−7.6	Semi-regular	131	M

Star	RA		Declination		Range	Type	Period (days)	Spectrum
	h	m	°	′				
RT Cygni	19	43.6	+48	47	6.0–13.1	Mira	190	M
SS Cygni	21	42.7	+43	35	7.7–12.4	Dwarf nova	50±	K+Pec
*CH Cygni	19	24.5	+50	14	5.6–9.0	Symbiotic	—	M+B
Chi Cygni	19	50.6	+32	55	3.3–14.2	Mira	408	S
R Delphini	20	14.9	+09	05	7.6–13.8	Mira	285	M
U Delphini	20	45.5	+18	05	5.6–7.5	Semi-regular	110?	M
EU Delphini	20	37.9	+18	16	5.8–6.9	Semi-regular	60	M
Beta Doradûs	05	33.6	−62	29	3.5–4.1	Cepheid	9.8	F–G
R Draconis	16	32.7	+66	45	6.7–13.2	Mira	246	M
T Eridani	03	55.2	−24	02	7.2–13.2	Mira	252	M
R Fornacis	02	29.3	−26	06	7.5–13.0	Mira	389	C
R Geminorum	07	07.4	+22	42	6.0–14.0	Mira	370	S
U Geminorum	07	55.1	+22	00	8.2–14.9	Dwarf nova	105±	Pec+M
Zeta Geminorum	07	04.1	+20	34	3.6–4.2	Cepheid	10.2	F–G
Eta Geminorum	06	14.9	+22	30	3.2–3.9	Semi-regular	233	M
S Gruis	22	26.1	−48	26	6.0–15.0	Mira	402	M
S Herculis	16	51.9	+14	56	6.4–13.8	Mira	307	M
U Herculis	16	25.8	+18	54	6.4–13.4	Mira	406	M
Alpha Herculis	17	14.6	+14	23	2.7–4.0	Semi-regular	—	M
68, u Herculis	17	17.3	+33	06	4.7–5.4	Algol	2.1	B+B
R Horologii	02	53.9	−49	53	4.7–14.3	Mira	408	M
U Horologii	03	52.8	−45	50	6–14	Mira	348	M
R Hydrae	13	29.7	−23	17	3.5–10.9	Mira	389	M
U Hydrae	10	37.6	−13	23	4.3–6.5	Semi-regular	450?	C
VW Hydri	04	09.1	−71	18	8.4–14.4	Dwarf nova	27±	Pec
R Leonis	09	47.6	+11	26	4.4–11.3	Mira	310	M
R Leonis Minoris	09	45.6	+34	31	6.3–13.2	Mira	372	M
R Leporis	04	59.6	−14	48	5.5–11.7	Mira	427	C
Y Librae	15	11.7	−06	01	7.6–14.7	Mira	276	M
RS Librae	15	24.3	−22	55	7.0–13.0	Mira	218	M
Delta Librae	15	01.0	−08	31	4.9–5.9	Algol	2.3	A
R Lyncis	07	01.3	+55	20	7.2–14.3	Mira	379	S
R Lyrae	18	55.3	+43	57	3.9–5.0	Semi-regular	46?	M
RR Lyrae	19	25.5	+42	47	7.1–8.1	RR Lyrae	0.6	A–F
Beta Lyrae	18	50.1	+33	22	3.3–4.4	Eclipsing	12.9	B
U Microscopii	20	29.2	−40	25	7.0–14.4	Mira	334	M
*U Monocerotis	07	30.8	−09	47	5.9–7.8	RV Tauri	91	F–K
V Monocerotis	06	22.7	−02	12	6.0–13.9	Mira	340	M
R Normae	15	36.0	−49	30	6.5–13.9	Mira	508	M
T Normae	15	44.1	−54	59	6.2–13.6	Mira	241	M

Star	RA		Declination		Range	Type	Period	Spectrum
	h	m	°	′			(days)	
R Octantis	05	26.1	−86	23	6.3−13.2	Mira	405	M
S Octantis	18	08.7	−86	48	7.2−14.0	Mira	259	M
V Ophiuchi	16	26.7	−12	26	7.3−11.6	Mira	297	C
X Ophiuchi	18	38.3	+08	50	5.9−9.2	Mira	329	M
RS Ophiuchi	17	50.2	−06	43	4.3−12.5	Recurrent nova	—	OB+M
U Orionis	05	55.8	+20	10	4.8−13.0	Mira	368	M
W Orionis	05	05.4	+01	11	5.9−7.7	Semi-regular	212	C
Alpha Orionis	05	55.2	+07	24	0.0−1.3	Semi-regular	2335	M
S Pavonis	19	55.2	−59	12	6.6−10.4	Semi-regular	381	M
Kappa Pavonis	18	56.9	−67	14	3.9−4.8	W Virginis	9.1	G
R Pegasi	23	06.8	+10	33	6.9−13.8	Mira	378	M
X Persei	03	55.4	+31	03	6.0−7.0	Gamma Cas.	—	O9.5
Beta Persei	03	08.2	+40	57	2.1−3.4	Algol	2.9	B
Zeta Phoenicis	01	08.4	−55	15	3.9−4.4	Algol	1.7	B+B
R Pictoris	04	46.2	−49	15	6.4−10.1	Semi-regular	171	M
RS Puppis	08	13.1	−34	35	6.5−7.7	Cepheid	41.4	F−G
L^2 Puppis	07	13.5	−44	39	2.6−6.2	Semi-regular	141	M
T Pyxidis	09	04.7	−32	23	6.5−15.3	Recurrent nova	7000±	Pec
U Sagittae	19	18.8	+19	37	6.5−9.3	Algol	3.4	B+G
WZ Sagittae	20	07.6	+17	42	7.0−15.5	Dwarf nova	1900±	A
R Sagittarii	19	16.7	−19	18	6.7−12.8	Mira	270	M
RR Sagittarii	19	55.9	−29	11	5.4−14.0	Mira	336	M
RT Sagittarii	20	17.7	−39	07	6.0−14.1	Mira	306	M
RU Sagittarii	19	58.7	−41	51	6.0−13.8	Mira	240	M
RY Sagittarii	19	16.5	−33	31	5.8−14.0	R Coronae Bor.	—	G
RR Scorpii	16	56.6	−30	35	5.0−12.4	Mira	281	M
RS Scorpii	16	55.6	−45	06	6.2−13.0	Mira	320	M
RT Scorpii	17	03.5	−36	55	7.0−15.2	Mira	449	S
Delta Scorpii	16	00.3	−22	37	1.6−2.3	Irregular	—	B
S Sculptoris	00	15.4	−32	03	5.5−13.6	Mira	363	M
R Scuti	18	47.5	−05	42	4.2−8.6	RV Tauri	146	G−K
R Serpentis	15	50.7	+15	08	5.2−14.4	Mira	356	M
S Serpentis	15	21.7	+14	19	7.0−14.1	Mira	372	M
T Tauri	04	22.0	+19	32	9.3−13.5	T Tauri	—	F−K
SU Tauri	05	49.1	+19	04	9.1−16.9	R Coronae Bor.	—	G
Lambda Tauri	04	00.7	+12	29	3.4−3.9	Algol	4.0	B+A
R Trianguli	02	37.0	+34	16	5.4−12.6	Mira	267	M
R Ursae Majoris	10	44.6	+68	47	6.5−13.7	Mira	302	M
T Ursae Majoris	12	36.4	+59	29	6.6−13.5	Mira	257	M
*Z Ursae Majoris	11	56.5	+57	52	6.2−9.4	Semi-regular	196	M

Star	RA		Declination		Range	Type	Period	Spectrum
	h	m	°	′			(days)	
*RY Ursae Majoris	12	20.5	+61	19	6.7–8.3	Semi-regular	310?	M
U Ursae Minoris	14	17.3	+66	48	7.1–13.0	Mira	331	M
R Virginis	12	38.5	+06	59	6.1–12.1	Mira	146	M
S Virginis	13	33.0	−07	12	6.3–13.2	Mira	375	M
SS Virginis	12	25.3	+00	48	6.0–9.6	Semi-regular	364	C
R Vulpeculae	21	04.4	+23	49	7.0–14.3	Mira	137	M
Z Vulpeculae	19	21.7	+25	34	7.3–8.9	Algol	2.5	B+A

V CANUM VENATICORUM   13h 19.5m +45° 32' (2000)

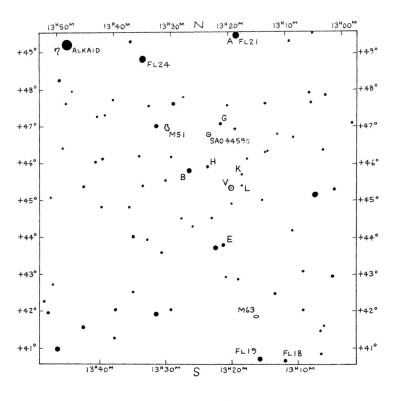

A 5.1	H 7.8
B 5.9	K 8.4
E 6.5	L 8.6
G 7.1	

o (MIRA) CETI   02h 19.3m −02° 59′ (2000)

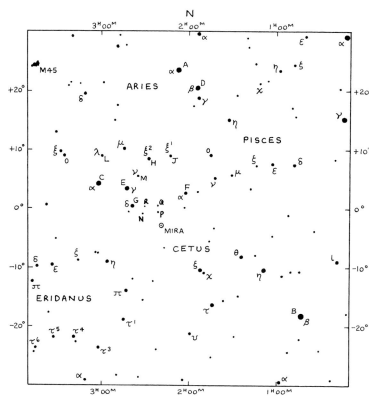

A	2.2	J	4.4
B	2.4	L	4.9
C	2.7	M	5.1
D	3.0	N	5.4
E	3.6	P	5.5
F	3.8	Q	5.7
G	4.1	R	6.1
H	4.3		

R CORONAE BOREALIS   15h 48.6m +28° 09′ (2000)

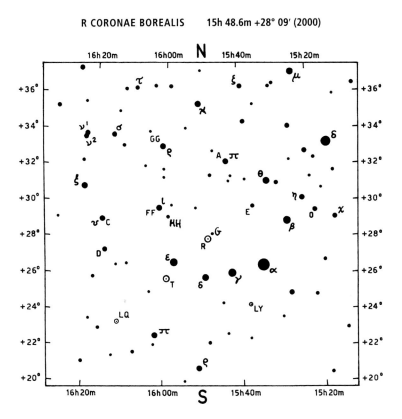

FF	5.0	C	5.8
GG	5.4	D	6.2
A	5.6	E	6.5
		HH	7.1
		G	7.4

CH CYGNI   19h 24.5m +50° 14′ (2000)

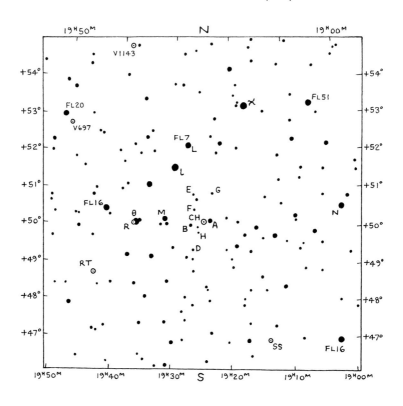

N 5.4	D 8.0
M 5.5	E 8.1
L 5.8	F 8.5
A 6.5	G 8.5
B 7.4	H 9.2

U MONOCEROTIS   07h 30.8m −09° 47' (2000)

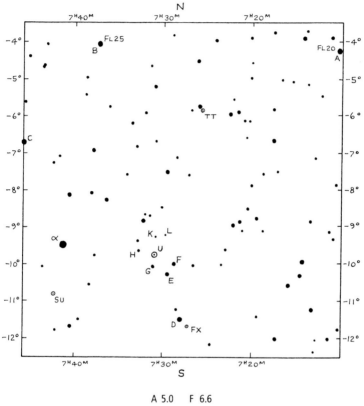

A 5.0	F 6.6
B 5.2	G 7.0
C 5.7	H 7.5
D 5.9	K 7.8
E 6.0	L 8.0

RY URSAE MAJORIS   12h 20.5m +61° 19' (2000)
Z URSAE MAJORIS   11h 56.5m +57° 52' (2000)

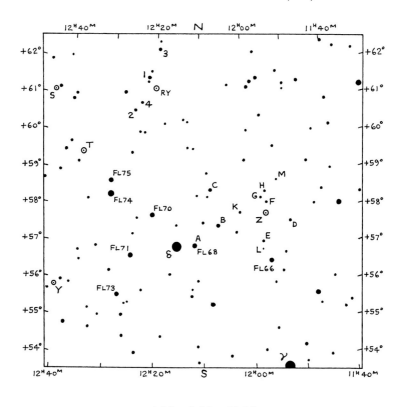

A 6.5	F 8.6	M 9.1
B 7.2	G 8.7	1 6.9
C 7.6	H 8.8	2 7.4
D 8.0	K 8.9	3 7.7
E 8.3	L 9.0	4 7.8

Mira Stars: Maxima, 2010

JOHN ISLES

Below are the predicted dates of maxima for Mira stars that reach magnitude 7.5 or brighter at an average maximum. Individual maxima can in some cases be brighter or fainter than average by a magnitude or more, and all dates are only approximate. The positions, extreme ranges and mean periods of these stars can be found in the preceding list of interesting variable stars.

Star	Mean Magnitude at Maximum	Dates of Maxima
R Andromedae	6.9	1 Oct
W Andromedae	7.4	4 Aug
R Aquarii	6.5	4 Jan
R Aquilae	6.1	3 Jun
R Boötis	7.2	8 Mar, 17 Oct
R Cancri	6.8	24 Aug
S Canis Minoris	7.5	30 Oct
R Carinae	4.6	24 Jan, 28 Nov
S Carinae	5.7	4 Apr, 31 Aug
R Cassiopeiae	7.0	21 Mar
R Centauri	5.8	15 Apr (secondary maximum, mean mag. 6.0)
T Cephei	6.0	24 Jan
U Ceti	7.5	15 Mar, 5 Nov
Omicron Ceti	3.4	16 Oct
T Columbae	7.5	30 May
S Coronae Borealis	7.3	28 Sep
V Coronae Borealis	7.5	14 Jan
R Corvi	7.5	27 Oct
R Cygni	7.5	11 Feb

Star	Mean Magnitude at Maximum	Dates of Maxima
U Cygni	7.2	25 Jan
RT Cygni	7.3	29 Jan, 7 Aug
R Geminorum	7.1	30 Dec
R Horologii	6.0	9 Feb
R Hydrae	4.5	27 Nov
R Leonis	5.8	20 Jul
R Leonis Minoris	7.1	16 Jan
R Leporis	6.8	14 Jan
RS Librae	7.5	8 Feb, 13 Sep
V Monocerotis	7.0	29 Apr
T Normae	7.4	2 Apr, 29 Nov
V Ophiuchi	7.5	8 Jan, 1 Nov
X Ophiuchi	6.8	24 Oct
U Orionis	6.3	23 Feb
R Sagittarii	7.3	31 Aug
RR Sagittarii	6.8	22 Oct
RT Sagittarii	7.0	2 May
RU Sagittarii	7.2	18 Jun
RR Scorpii	5.9	26 Sep
RS Scorpii	7.0	7 Feb, 23 Dec
S Sculptoris	6.7	1 Jan, 29 Dec
R Serpentis	6.9	22 Sep
R Trianguli	6.2	30 May
R Ursae Majoris	7.5	1 Feb, 29 Nov
R Virginis	6.9	24 May, 17 Oct
S Virginis	7.0	14 Feb

Some Interesting Double Stars

BOB ARGYLE

The positions, angles, and separations given below correspond to epoch 2010.0.

No.	RA		Declin-ation		Star	Magni-tudes	Separa-tion	PA	Cata-logue	Comments
	h	m	°	′			arcsec	°		
1	00	31.5	−62	58	β Tuc	4.4, 4.8	27.1	169	LCL 119	Both again difficult doubles.
2	00	49.1	+57	49	η Cas	3.4, 7.5	13.2	322	Σ60	Easy. Creamy, bluish. P = 480 years.
3	00	55.0	+23	38	36 And	6.0, 6.4	1.1	323	Σ73	P = 168 years. Both yellow. Slowly opening.
4	01	13.7	+07	35	ζ Psc	5.6, 6.5	23.1	63	Σ100	Yellow, reddish-white.
5	01	39.8	−56	12	p Eri	5.8, 5.8	11.7	188	Δ5	Period = 483 years.
6	01	53.5	+19	18	γ Ari	4.8, 4.8	7.5	1	Σ180	Very easy. Both white.
7	02	02.0	+02	46	α Psc	4.2, 5.1	1.8	265	Σ202	Binary, period = 933 years.
8	02	03.9	+42	20	γ And	2.3, 5.0	9.6	63	Σ205	Yellow, blue. Relatively fixed.
					γ2 And	5.1, 6.3	0.2	99	OΣ38	BC. Needs 30 cm. Closing.
9	02	29.1	+67	24	ι Cas AB	4.9, 6.9	2.6	229	Σ262	AB is long period binary. P = 620 years.
					ι Cas AC	4.9, 8.4	7.2	118		
10	02	33.8	−28	14	ω For	5.0, 7.7	10.8	245	HJ 3506	Common proper motion.

No.	RA		Declin-ation		Star	Magni-tudes	Separa-tion	PA	Cata-logue	Comments
	h	m	°	′			arcsec	°		
11	02	43.3	+03	14	γ Cet	3.5, 7.3	2.3	298	Σ299	Not too easy.
12	02	58.3	−40	18	θ Eri	3.4, 4.5	8.3	90	PZ 2	Both white.
13	02	59.2	+21	20	ε Ari	5.2, 5.5	1.4	208	Σ333	Closing slowly. P = 350 years? Both white.
14	03	00.9	+52	21	Σ331 Per	5.3, 6.7	12.0	85	−	Fixed.
15	03	12.1	−28	59	α For	4.0, 7.0	5.2	300	HJ 3555	P = 269 years. B variable?
16	03	48.6	−37	37	f Eri	4.8, 5.3	8.2	215	Δ16	Pale yellow. Fixed.
17	03	54.3	−02	57	32 Eri	4.8, 6.1	6.9	348	Σ470	Fixed.
18	04	32.0	+53	55	1 Cam	5.7, 6.8	10.3	308	Σ550	Fixed.
19	04	50.9	−53	28	ι Pic	5.6, 6.4	12.4	58	Δ18	Good object for small apertures. Fixed.
20	05	13.2	−12	56	κ Lep	4.5, 7.4	2.0	357	Σ661	Visible in 7.5 cm. Slowly closing.
21	05	14.5	−08	12	β Ori	0.1, 6.8	9.5	204	Σ668	Companion once thought to be close double.
22	05	21.8	−24	46	41 Lep	5.4, 6.6	3.4	93	HJ 3752	Deep yellow pair in a rich field.
23	05	24.5	−02	24	η Ori	3.8, 4.8	1.7	78	DA 5	Slow moving binary.
24	05	35.1	+09	56	λ Ori	3.6, 5.5	4.3	44	Σ738	Fixed.
25	05	35.3	−05	23	θ Ori AB	6.7, 7.9	8.6	32	Σ748	Trapezium in M42.
					θ Ori CD	5.1, 6.7	13.4	61		
26	05	40.7	−01	57	ζ Ori	1.9, 4.0	2.6	166	Σ774	Can be split in 7.5 cm. Long period binary.
27	06	14.9	+22	30	η Gem	var, 6.5	1.6	254	β1008	Well seen with 20 cm. Primary orange.
28	06	46.2	+59	27	12 Lyn AB	5.4, 6.0	1.9	69	Σ948	AB is binary, P = 706 years.
					12 Lyn AC	5.4, 7.3	8.7	309		

No.	RA		Declin-ation		Star	Magni-tudes	Separa-tion	PA	Cata-logue	Comments
	h	m	°	′			arcsec	°		
29	07	08.7	−70	30	γ Vol	3.9, 5.8	14.1	298	Δ42	Very slow binary.
30	07	16.6	−23	19	h3945 CMa	4.8, 6.8	26.8	51	–	Contrasting colours. Yellow and blue.
31	07	20.1	+21	59	δ Gem	3.5, 8.2	5.6	227	Σ1066	Not too easy. Yellow, pale blue.
32	07	34.6	+31	53	α Gem	1.9, 2.9	4.6	57	Σ1110	Widening. Easy with 7.5 cm.
33	07	38.8	−26	48	κ Pup	4.5, 4.7	9.8	318	H III 27	Both white.
34	08	12.2	+17	39	ζ Cnc AB	5.6, 6.0	1.1	38	Σ1196	Period (AB) = 60 years. Near maximum separation.
					ζ Cnc AB-C	5.0, 6.2	5.9	68	Σ1196	Period (AB-C) = 1,150 years.
35	08	46.8	+06	25	ε Hyd	3.3, 6.8	2.9	305	Σ1273	PA slowly increasing. A is a very close pair.
36	09	18.8	+36	48	38 Lyn	3.9, 6.6	2.6	226	Σ1334	Almost fixed.
37	09	47.1	−65	04	μ Car	3.1, 6.1	5.0	129	RMK 11	Fixed. Fine in small telescopes.
38	10	20.0	+19	50	γ Leo	2.2, 3.5	4.6	126	Σ1424	Binary, period = 619 years. Both orange.
39	10	32.0	−45	04	s Vel	6.2, 6.5	13.5	218	PZ 3	Fixed.
40	10	46.8	−49	26	μ Vel	2.7, 6.4	2.6	56	R 155	P = 138 years. Near widest separation.
41	10	55.6	+24	45	54 Leo	4.5, 6.3	6.6	111	Σ1487	Slowly widening. Pale yellow and white.
42	11	18.2	+31	32	ξ UMa	4.3, 4.8	1.6	211	Σ1523	Binary, 60 years. Needs 7.5 cm.

No.	RA		Declin-ation		Star	Magni-tudes	Separa-tion	PA	Cata-logue	Comments
	h	m	°	′			arcsec	°		
43	11	23.9	+10	32	ι Leo	4.0, 6.7	2.0	100	Σ1536	Binary, period = 186 years.
44	11	32.3	−29	16	N Hya	5.8, 5.9	9.4	210	H III 96	Fixed.
45	12	14.0	−45	43	D Cen	5.6, 6.8	2.8	243	RMK 14	Orange and white. Closing.
46	12	26.6	−63	06	α Cru	1.4, 1.9	4.0	114	Δ252	Third star in a low power field.
47	12	41.5	−48	58	γ Cen	2.9, 2.9	0.4	324	HJ 4539	Period = 84 years. Closing. Both yellow.
48	12	41.7	−01	127	γ Vir	3.5, 3.5	1.4	24	Σ1670	Periastron in 2005. Now widening quickly.
49	12	46.3	−68	06	β Mus	3.7, 4.0	1.0	52	R 207	Both white. Closing slowly. P = 383 years.
50	12	54.6	−57	11	μ Cru	4.3, 5.3	34.9	17	Δ126	Fixed. Both white.
51	12	56.0	+38	19	α CVn	2.9, 5.5	19.3	229	Σ1692	Easy. Yellow, bluish.
52	13	22.6	−60	59	J Cen	4.6, 6.5	60.0	343	Δ133	Fixed. A is a close pair.
53	13	24.0	+54	56	ζ UMa	2.3, 4.0	14.4	152	Σ1744	Very easy. Naked-eye pair with Alcor.
54	13	51.8	−33	00	3 Cen	4.5, 6.0	7.7	102	H III 101	Both white. Closing slowly.
55	14	39.6	−60	50	α Cen	0.0, 1.2	6.8	245	RHD 1	Finest pair in the sky. P = 80 years. Closing.
56	14	41.1	+13	44	ζ Boo	4.5, 4.6	0.5	294	Σ1865	Both white. Closing – highly inclined orbit.
57	14	45.0	+27	04	ε Boo	2.5, 4.9	2.9	344	Σ1877	Yellow, blue. Fine pair.
58	14	46.0	−25	27	54 Hya	5.1, 7.1	8.3	122	H III 97	Closing slowly.

No.	RA		Declin-ation		Star	Magni-tudes	Separa-tion	PA	Cata-logue	Comments
	h	m	°	′			arcsec	°		
59	14	49.3	−14	09	μ Lib	5.8, 6.7	1.8	6	β106	Becoming wider. Fine in 7.5 cm.
60	14	51.4	+19	06	ξ Boo	4.7, 7.0	6.0	308	Σ1888	Fine contrast. Easy.
61	15	03.8	+47	39	44 Boo	5.3, 6.2	1.6	60	Σ1909	Period = 206 years. Beginning to close.
62	15	05.1	−47	03	π Lup	4.6, 4.7	1.7	66	HJ 4728	Widening.
63	15	18.5	−47	53	μ Lup AB	5.1, 5.2	1.1	300	HJ 4753	AB closing. Underobserved.
					μ Lup AC	4.4, 7.2	22.7	127	Δ180	AC almost fixed.
64	15	23.4	−59	19	γ Cir	5.1, 5.5	0.7	0	HJ 4757	Closing. Needs 20 cm. Long period binary.
65	15	34.8	+10	33	δ Ser	4.2, 5.2	4.0	173	Σ1954	Long period binary.
66	15	35.1	−41	10	γ Lup	3.5, 3.6	0.8	277	HJ 4786	Binary. Period = 190 years. Needs 20 cm.
67	15	56.9	−33	58	ξ Lup	5.3, 5.8	10.2	49	PZ 4	Fixed.
68	16	14.7	+33	52	σ CrB	5.6, 6.6	7.2	237	Σ2032	Long period binary. Both white.
69	16	29.4	−26	26	α Sco	1.2, 5.4	2.6	277	GNT 1	Red, green. Difficult from mid-northern latitudes.
70	16	30.9	+01	59	λ Oph	4.2, 5.2	1.4	37	Σ2055	P = 129 years. Fairly difficult in small apertures.
71	16	41.3	+31	36	ζ Her	2.9, 5.5	1.1	178	Σ2084	Period = 34 years. Now widening. Needs 20 cm.
72	17	05.3	+54	28	μ Dra	5.7, 5.7	2.4	6	Σ2130	Period = 672 years.
73	17	14.6	+14	24	α Her	var, 5.4	4.6	103	Σ2140	Red, green. Long period binary.

No.	RA		Declin-ation		Star	Magni-tudes	Separa-tion	PA	Cata-logue	Comments
	h	m	°	′			arcsec	°		
74	17	15.3	−26	35	36 Oph	5.1, 5.1	5.0	143	SHJ 243	Period = 471 years.
75	17	23.7	+37	08	ρ Her	4.6, 5.6	4.1	319	Σ2161	Slowly widening.
76	17	26.9	−45	51	HJ 4949 AB	5.6, 6.5	2.1	251	HJ 4949	Beautiful coarse triple. All white.
					Δ 216 AC	7.1	105.0	310		
77	18	01.5	+21	36	95 Her	5.0, 5.1	6.5	257	Σ2264	Colours thought variable in C19.
78	18	05.5	+02	30	70 Oph	4.2, 6.0	5.7	132	Σ2272	Opening. Easy in 7.5 cm.
79	18	06.8	−43	25	h5014 CrA	5.7, 5.7	1.7	2	–	Period = 450 years. Needs 10 cm.
80	18	25.4	−20	33	21 Sgr	5.0, 7.4	1.7	279	JC 6	Slowly closing binary, orange and green.
81	18	35.9	+16	58	OΣ358 Her	6.8, 7.0	1.5	149	–	Period = 380 years.
82	18	44.3	+39	40	ε^1 Lyr	5.0, 6.1	2.4	348	Σ2382	Quadruple system with ε^2. Both pairs
83	18	44.3	+39	40	ε^2 Lyr	5.2, 5.5	2.4	78	Σ2383	visible in 7.5 cm.
84	18	56.2	+04	12	θ Ser	4.5, 5.4	22.4	104	Σ2417	Fixed. Very easy.
85	19	06.4	−37	04	γ CrA	4.8, 5.1	1.3	13	HJ 5084	Beautiful pair. Period = 122 years.
86	19	30.7	+27	58	β Cyg AB	3.1, 5.1	34.3	54	Σ I 43	Glorious. Yellow, blue-greenish.
					β Cyg Aa	3.1, 5.2	0.4	98	MCA 55	Aa. Period = 97 years. Closing.
87	19	45.0	+45	08	δ Cyg	2.9, 6.3	2.7	220	Σ2579	Slowly widening. Period = 780 years.
88	19	48.2	+70	16	ε Dra	3.8, 7.4	3.3	20	Σ2603	Slow binary.
89	19	54.6	−08	14	57 Aql	5.7, 6.4	36.0	170	Σ2594	Easy pair. Contrasting colours.

No.	RA		Declin- ation		Star	Magni- tudes	Separa- tion	PA	Cata- logue	Comments
	h	m	°	′			arcsec	°		
90	20	46.7	+16	07	γ Del	4.5, 5.5	9.1	265	Σ2727	Easy. Yellowish. Long period binary.
91	20	59.1	+04	18	ε Equ AB	6.0, 6.3	0.5	284	Σ2737	Fine triple. AB is closing.
					ε Equ AC	6.0, 7.1	10.3	66		
92	21	06.9	+38	45	61 Cyg	5.2, 6.0	31.3	151	Σ2758	Nearby binary. Both orange. Period = 659 years.
93	21	19.9	−53	27	θ Ind	4.5, 7.0	7.0	271	HJ 5258	Pale yellow and reddish. Long period binary.
94	21	44.1	+28	45	μ Cyg	4.8, 6.1	1.7	316	Σ2822	Period = 789 years.
95	22	03.8	+64	37	ξ Cep	4.4, 6.5	8.3	274	Σ2863	White and blue. Long period binary.
96	22	14.3	−21	04	41 Aqr	5.6, 6.7	5.1	113	H N 56	Yellowish and purple?
97	22	26.6	−16	45	53 Aqr	6.4, 6.6	1.3	35	SHJ 345	Long period binary; periastron in 2023.
98	22	28.8	−00	01	ζ Aqr	4.3, 4.5	2.2	173	Σ2909	Period = 587 years. Slowly widening.
99	23	19.1	−13	28	94 Aqr	5.3, 7.0	12.3	351	Σ2988	Yellow and orange. Probable binary.
100	23	59.5	+33	43	Σ3050 And	6.6, 6.6	2.1	336	–	Period = 350 years.

Some Interesting Nebulae, Clusters, and Galaxies

Object	RA		Declina-tion		Remarks
	h	m	°	′	
M31 Andromedae	00	40.7	+41	05	Andromeda Galaxy, visible to naked eye.
H VIII 78 Cassiopeiae	00	41.3	+61	36	Fine cluster, between Gamma and Kappa Cassiopeiae.
M33 Trianguli	01	31.8	+30	28	Spiral. Difficult with small apertures.
H VI 33−4 Persei, C14	02	18.3	+56	59	Double cluster; Sword-handle.
Δ142 Doradus	05	39.1	−69	09	Looped nebula round 30 Doradus. Naked eye. In Large Magellanic Cloud.
M1 Tauri	05	32.3	+22	00	Crab Nebula, near Zeta Tauri.
M42 Orionis	05	33.4	−05	24	Orion Nebula. Contains the famous Trapezium, Theta Orionis.
M35 Geminorum	06	06.5	+24	21	Open cluster near Eta Geminorum.
H VII 2 Monocerotis, C50	06	30.7	+04	53	Open cluster, just visible to naked eye.
M41 Canis Majoris	06	45.5	−20	42	Open cluster, just visible to naked eye.
M47 Puppis	07	34.3	−14	22	Mag. 5.2. Loose cluster.
H IV 64 Puppis	07	39.6	−18	05	Bright planetary in rich neighbourhood.
M46 Puppis	07	39.5	−14	42	Open cluster.
M44 Cancri	08	38	+20	07	Praesepe. Open cluster near Delta Cancri. Visible to naked eye.
M97 Ursae Majoris	11	12.6	+55	13	Owl Nebula, diameter 3′. Planetary.
Kappa Crucis, C94	12	50.7	−60	05	'Jewel Box'; open cluster, with stars of contrasting colours.
M3 Can. Ven.	13	40.6	+28	34	Bright globular.
Omega Centauri, C80	13	23.7	−47	03	Finest of all globulars. Easy with naked eye.
M80 Scorpii	16	14.9	−22	53	Globular, between Antares and Beta Scorpii.
M4 Scorpii	16	21.5	−26	26	Open cluster close to Antares.

Object	RA		Declina-tion		Remarks
	h	m	°	′	
M13 Herculis	16	40	+36	31	Globular. Just visible to naked eye.
M92 Herculis	16	16.1	+43	11	Globular. Between Iota and Eta Herculis.
M6 Scorpii	17	36.8	−32	11	Open cluster; naked eye.
M7 Scorpii	17	50.6	−34	48	Very bright open cluster; naked eye.
M23 Sagittarii	17	54.8	−19	01	Open cluster nearly 50′ in diameter.
H IV 37 Draconis, C6	17	58.6	+66	38	Bright planetary.
M8 Sagittarii	18	01.4	−24	23	Lagoon Nebula. Gaseous. Just visible with naked eye.
NGC 6572 Ophiuchi	18	10.9	+06	50	Bright planetary, between Beta Ophiuchi and Zeta Aquilae.
M17 Sagittarii	18	18.8	−16	12	Omega Nebula. Gaseous. Large and bright.
M11 Scuti	18	49.0	−06	19	Wild Duck. Bright open cluster.
M57 Lyrae	18	52.6	+32	59	Ring Nebula. Brightest of planetaries.
M27 Vulpeculae	19	58.1	+22	37	Dumb-bell Nebula, near Gamma Sagittae.
H IV 1 Aquarii, C55	21	02.1	−11	31	Bright planetary, near Nu Aquarii.
M15 Pegasi	21	28.3	+12	01	Bright globular, near Epsilon Pegasi.
M39 Cygni	21	31.0	+48	17	Open cluster between Deneb and Alpha Lacertae. Well seen with low powers.

(M = Messier number; NGC = New General Catalogue number; C = Caldwell number.)

Our Contributors

Paul G. Abel is a lunar and planetary amateur astronomer who makes all of his observations visually using drawings to capture the details of the Moon and planets. He has rather a soft spot for Saturn and is Assistant Director of the Saturn Section of the British Astronomical Association. Paul frequently makes the journey to Selsey to use one of his favourite telescopes – Patrick Moore's 12.5-inch reflector – and to eat powerful curries with Patrick. By profession Paul is a theoretical physicist whose research interests lie in General Relativity, Hawking Radiation and Quantum Field Theory in curved space-times. He currently teaches Mathematics in the Department of Physics and Astronomy at the University of Leicester.

Dr David A. Rothery chairs courses in planetary science and about volcanoes, earthquakes and tsunamis at the Open University. He has been interested in space since childhood, and made a research career as a geologist by using satellite images to study the Earth's surface. He was a geologist on the ill-fated Beagle 2 project, and is now 'UK Lead Scientist' on the Mercury Imaging X-ray Spectrometer. He has served on various research councils and space agency advisory boards.

Martin Mobberley is one of the UK's most active imagers of comets, planets, asteroids, variable stars, novae and supernovae and served as President of the British Astronomical Association from 1997 to 1999. In 2000 he was awarded the Association's Walter Goodacre Award. He is the author of six astronomy books published by Springer as well as three children's space exploration books published by Top That Publishing.

Professor Fred Watson is Astronomer-in-Charge of the Anglo–Australian Observatory at Coonabarabran in north-western New South Wales, and a well-known broadcaster on Australian radio. He is a regular contributor to the *Yearbook of Astronomy*, and his recent books include *Universe* (for which he was Chief Consultant), *Stargazer: The Life and Times of the Telescope*, and *Why is Uranus upside down? And Other Questions about the Universe*. In 2006 he was awarded the Australian Government Eureka Prize for Promoting Understanding of Science. Visit Fred's website at http://fredwatson.com.au/.

Dr David M. Harland gained his BSc in astronomy in 1977 and a doctorate in computational science. Subsequently, he has taught computer science, worked in industry and managed academic research. In 1995 he 'retired' and has since published many books on space themes.

Iain Nicolson, formerly Principal Lecturer in Astronomy at the University of Hertfordshire, is a writer and lecturer in the fields of astronomy and space science. A Contributing Consultant to the magazine *Astronomy Now*, he has been a frequent contributor to BBC Television's *The Sky at Night*. He is author or co-author of more than twenty books, the most recent of which, *Dark Side of the Universe*, was published in April 2007 by Canopus Publishing Limited.

Professor Greg Parker is Professor of Photonics in the School of Electronics and Computer Science at the University of Southampton. By day he carries out research into optical nanodevices called photonic quasicrystals, and by night he is a keen amateur astrophotographer. Greg has published over 120 refereed journal papers in solid-state physics and optics, and two books (a solid-state physics textbook and an astrophotography book), and has filed over a dozen patents. He is a C.Eng., C.Phys. and F.Inst.P., and is currently building his own mini-Wasp array at the New Forest Observatory http://www.newforest observatory.com/. He is married with one son, a dog, a cat, a pond full of fish, and more computers and optical bits and pieces than you can throw a stick at.

Dr Allan Chapman, of Wadham College, Oxford, is probably Britain's leading authority on the history of astronomy. He has published many research papers and several books, as well as numerous popular accounts. He is a frequent and welcome contributor to the *Yearbook*.

Astronomical Societies in the British Isles

British Astronomical Association
Assistant Secretary: Burlington House, Piccadilly, London W1V 9AG.
Meetings: Lecture Hall of Scientific Societies, Civil Service Commission Building, 23 Savile Row, London W1. Last Wednesday each month (Oct.–June), 5 p.m. and some Saturday afternoons.

Association for Astronomy Education
Secretary: Teresa Grafton, The Association for Astronomy Education, c/o The Royal Astronomical Society, Burlington House, Piccadilly, London W1V 0NL.

Astronomical Society of Edinburgh
Secretary: Graham Rule, 105/19 Causewayside, Edinburgh EH9 1QG.
Website: www.roe.ac.uk/asewww/; *Email:* asewww@roe.ac.uk
Meetings: City Observatory, Calton Hill, Edinburgh. 1st Friday each month, 8 p.m.

Astronomical Society of Glasgow
Secretary: Mr David Degan, 5 Hillside Avenue, Alexandria, Dunbartonshire G83 0BB.
Website: www.astronomicalsocietyofglasgow.org.uk
Meetings: Royal College, University of Strathclyde, Montrose Street, Glasgow. 3rd Thursday each month, Sept.–Apr., 7.30 p.m.

Astronomical Society of Haringey
Secretary: Jerry Workman, 91 Greenslade Road, Barking, Essex IG11 9XF.
Meetings: Palm Court, Alexandra Palace, 3rd Wednesday each month, 8 p.m.

Astronomy Ireland
Secretary: Tony Ryan, PO Box 2888, Dublin 1, Eire.
Website: www.astronomy.ie; *Email:* info@astronomy.ie
Meetings: 2nd Monday of each month. Telescope meetings every clear Saturday.

Federation of Astronomical Societies
Secretary: Clive Down, 10 Glan-y-Llyn, North Cornelly, Bridgend, County Borough CF33 4EF.
Email: clivedown@btinternet.com

Junior Astronomical Society of Ireland
Secretary: K. Nolan, 5 St Patrick's Crescent, Rathcoole, Co. Dublin.
Meetings: The Royal Dublin Society, Ballsbridge, Dublin 4. Monthly.

Society for Popular Astronomy
Secretary: Guy Fennimore, 36 Fairway, Keyworth, Nottingham NG12 5DU.
Website: www.popastro.com; *Email:* SPAstronomy@aol.com
Meetings: Last Saturday in Jan., Apr., July, Oct., 2.30 p.m. in London.

Webb Society
Treasurer/Membership Secretary: Steve Rayner, 10 Meon Close, Tadley RG26 4HN.

Aberdeen and District Astronomical Society
Secretary: Ian C. Giddings, 95 Brentfield Circle, Ellon, Aberdeenshire AB41 9DB.
Meetings: Robert Gordon's Institute of Technology, St Andrew's Street, Aberdeen.
Fridays, 7.30 p.m.

Abingdon Astronomical Society (was **Fitzharry's Astronomical Society**)
Secretary: Chris Holt, 9 Rutherford Close, Abingdon, Oxon OX14 2AT.
Website: www.abingdonastro.org.uk; *Email:* info@abingdonastro.co.uk
Meetings: All Saints' Methodist Church Hall, Dorchester Crescent, Abingdon, Oxon.
2nd Monday Sept.–June, 8 p.m. and additional beginners' meetings and observing
evenings as advertised.

Altrincham and District Astronomical Society
Secretary: Derek McComiskey, 33 Tottenham Drive, Manchester M23 9WH.
Meetings: Timperley Village Club. 1st Friday Sept.–June, 8 p.m.

Andover Astronomical Society
Secretary: Mrs S. Fisher, Staddlestones, Aughton, Kingston, Marlborough, Wiltshire
SN8 3SA.
Meetings: Grately Village Hall. 3rd Thursday each month, 7.30 p.m.

Astra Astronomy Section
Secretary: c/o Duncan Lunan, Flat 65, Dalraida House, 56 Blythswood Court,
Anderston, Glasgow G2 7PE.
Meetings: Airdrie Arts Centre, Anderson Street, Airdrie. Weekly.

Astrodome Mobile School Planetarium
Contact: Peter J. Golding, 53 City Way, Rochester, Kent ME1 2AX.
Website: www.astrodome.clara.co.uk; *Email:* astrodome@clara.co.uk

Aylesbury Astronomical Society
Secretary: Alan Smith, 182 Marley Fields, Leighton Buzzard, Bedfordshire LU7 8WN.
Meetings: 1st Monday in month at 8 p.m., venue in Aylesbury area. Details from
Secretary.

Bassetlaw Astronomical Society
Secretary: Andrew Patton, 58 Holding, Worksop, Notts S81 0TD.
Meetings: Rhodesia Village Hall, Rhodesia, Worksop, Notts. 2nd and 4th Tuesdays of
month at 7.45 p.m.

Batley & Spenborough Astronomical Society
Secretary: Robert Morton, 22 Links Avenue, Cleckheaton, West Yorks BD19 4EG.
Meetings: Milner K. Ford Observatory, Wilton Park, Batley. Every Thursday, 8 p.m.

Bedford Astronomical Society
Secretary: Mrs L. Harrington, 24 Swallowfield, Wyboston, Bedfordshire MK44 3AE.
Website: www.observer1.freeserve.co.uk/bashome.html
Meetings: Bedford School, Burnaby Rd, Bedford. Last Wednesday each month.

Bingham & Brooks Space Organization
Secretary: N. Bingham, 15 Hickmore's Lane, Lindfield, West Sussex.

Birmingham Astronomical Society
Contact: P. Bolas, 4 Moat Bank, Bretby, Burton-on-Trent DE15 0QJ.
Website: www.birmingham-astronomical.co.uk; *Email:* pbolas@aol.com
Meetings: Room 146, Aston University. Last Tuesday of month. Sept.–June (except
Dec., moved to 1st week in Jan.).

Blackburn Leisure Astronomy Section
Secretary: Mr H. Murphy, 20 Princess Way, Beverley, East Yorkshire HU17 8PD.
Meetings: Blackburn Leisure Welfare. Mondays, 8 p.m.

Blackpool & District Astronomical Society
Secretary: Terry Devon, 30 Victory Road, Blackpool, Lancashire FY1 3JT.
Website: www.blackpoolastronomy.org.uk; *Email:* info@blackpoolastronomy.org.uk
Meetings: St Kentigern's Social Centre, Blackpool. 1st Wednesday of the month,
7.45 p.m.

Bolton Astronomical Society
Secretary: Peter Miskiw, 9 Hedley Street, Bolton, Lancashire BL1 3LE.
Meetings: Ladybridge Community Centre, Bolton. 1st and 3rd Tuesdays Sept.–May,
7.30 p.m.

Border Astronomy Society
Secretary: David Pettitt, 14 Sharp Grove, Carlisle, Cumbria CA2 5QR.
Website: www.members.aol.com/P3pub/page8.html
Email: davidpettitt@supanet.com
Meetings: The Observatory, Trinity School, Carlisle. Alternate Thursdays, 7.30 p.m.,
Sept.–May.

Boston Astronomers
Secretary: Mrs Lorraine Money, 18 College Park, Horncastle, Lincolnshire LN9 6RE.
Meetings: Blackfriars Arts Centre, Boston. 2nd Monday each month, 7.30 p.m.

Bradford Astronomical Society
Contact: Mrs J. Hilary Knaggs, 6 Meadow View, Wyke, Bradford BD12 9LA.
Website: www.bradford-astro.freeserve.co.uk/index.htm
Meetings: Eccleshill Library, Bradford. Alternate Mondays, 7.30 p.m.

Braintree, Halstead & District Astronomical Society
Secretary: Mr J. R. Green, 70 Dorothy Sayers Drive, Witham, Essex CM8 2LU.
Meetings: BT Social Club Hall, Witham Telephone Exchange. 3rd Thursday each
month, 8 p.m.

Breckland Astronomical Society (was **Great Ellingham and District Astronomy Club**)
Contact: Martin Wolton, Willowbeck House, Pulham St Mary, Norfolk IP21 4QS.
Meetings: Great Ellingham Recreation Centre, Watton Road (B1077), Great
Ellingham, 2nd Friday each month, 7.15 p.m.

Bridgend Astronomical Society
Secretary: Clive Down, 10 Glan-y-Llyn, Broadlands, North Cornelly, Bridgend
County CF33 4EF.
Email: clivedown@btinternet.com
Meetings: Bridgend Bowls Centre, Bridgend. 2nd Friday, monthly, 7.30 p.m.

Bridgwater Astronomical Society
Secretary: Mr G. MacKenzie, Watergore Cottage, Watergore, South Petherton,
Somerset TA13 5JQ.
Website: www.ourworld.compuserve.com/hompages/dbown/Bwastro.htm
Meetings: Room D10, Bridgwater College, Bath Road Centre, Bridgwater. 2nd
Wednesday each month, Sept.–June.

Bridport Astronomical Society
Secretary: Mr G.J. Lodder, 3 The Green, Walditch, Bridport, Dorset DT6 4LB.
Meetings: Walditch Village Hall, Bridport. 1st Sunday each month, 7.30 p.m.

Brighton Astronomical and Scientific Society
Secretary: Ms T. Fearn, 38 Woodlands Close, Peacehaven, East Sussex BN10 7SF.
Meetings: St John's Church Hall, Hove. 1st Tuesday each month, 7.30 p.m.

Bristol Astronomical Society
Secretary: Dr John Pickard, 'Fielding', Easter Compton, Bristol BS35 5SJ.
Meetings: Frank Lecture Theatre, University of Bristol Physics Dept., alternate Fridays in term time, and Westbury Park Methodist Church Rooms, North View, other Fridays.

Callington Community Astronomy Group
Secretary: Beccy Watson. *Tel:* 07891 573786
Email: enquiries@callington-astro.org.uk
Website: www.callington-astro.org.uk
Meetings: Callington Space Centre, Callington Community College, Launceston Road, Callington, Cornwall PL17 7DR. 1st Friday of each month, 7.30 p.m., Sept.–June.

Cambridge Astronomical Society
Secretary: Brian Lister, 80 Ramsden Square, Cambridge CB4 2BL.
Meetings: Institute of Astronomy, Madingley Road. 3rd Friday each month.

Cardiff Astronomical Society
Secretary: D.W.S. Powell, 1 Tal-y-Bont Road, Ely, Cardiff CF5 5EU.
Meetings: Dept. of Physics and Astronomy, University of Wales, Newport Road, Cardiff. Alternate Thursdays, 8 p.m.

Castle Point Astronomy Club
Secretary: Andrew Turner, 3 Canewdon Hall Close, Canewdon, Rochford, Essex SS4 3PY.
Meetings: St Michael's Church Hall, Daws Heath. Wednesdays, 8 p.m.

Chelmsford Astronomers
Secretary: Brendan Clark, 5 Borda Close, Chelmsford, Essex.
Meetings: Once a month.

Chester Astronomical Society
Secretary: John Gilmour, 2 Thomas Brassey Close, Chester CH2 3AE.
Tel.: 07974 948278
Email: john_gilmour@ouvip.com
Website: www.manastro.co.uk/nwgas/chester/
Meetings: Burley Memorial Hall, Waverton, near Chester. Last Wednesday of each month except August and December at 7.30 p.m.

Chester Society of Natural Science, Literature and Art
Secretary: Paul Braid, 'White Wing', 38 Bryn Avenue, Old Colwyn, Colwyn Bay LL29 8AH.
Email: p.braid@virgin.net
Meetings: Once a month.

Chesterfield Astronomical Society
President: Mr D. Blackburn, 71 Middlecroft Road, Stavely, Chesterfield, Derbyshire S41 3XG. Tel: 07909 570754.
Website: www.chesterfield-as.org.uk
Meetings: Barnet Observatory, Newbold, each Friday.

Clacton & District Astronomical Society
Secretary: C. L. Haskell, 105 London Road, Clacton-on-Sea, Essex.

Cleethorpes & District Astronomical Society
Secretary: C. Illingworth, 38 Shaw Drive, Grimsby, South Humberside.
Meetings: Beacon Hill Observatory, Cleethorpes. 1st Wednesday each month.

Cleveland & Darlington Astronomical Society
Contact: Dr John McCue, 40 Bradbury Rd., Stockton-on-Tees, Cleveland TS20 1LE.
Meetings: Grindon Parish Hall, Thorpe Thewles, near Stockton-on-Tees. 2nd Friday, monthly.

Cork Astronomy Club
Secretary: Charles Coughlan, 12 Forest Ridge Crescent, Wilton, Cork, Eire.
Meetings: 1st Monday, Sept.–May (except bank holidays).

Cornwall Astronomical Society
Secretary: J.M. Harvey, 1 Tregunna Close, Porthleven, Cornwall TR13 9LW.
Meetings: Godolphin Club, Wendron Street, Helston, Cornwall. 2nd and 4th Thursday of each month, 7.30 for 8 p.m.

Cotswold Astronomical Society
Secretary: Rod Salisbury, Grove House, Christchurch Road, Cheltenham, Gloucestershire GL50 2PN.
Website: www.members.nbci.com/CotswoldAS
Meetings: Shurdington Church Hall, School Lane, Shurdington, Cheltenham. 2nd Saturday each month, 8 p.m.

Coventry & Warwickshire Astronomical Society
Secretary: Steve Payne, 68 Stonebury Avenue, Eastern Green, Coventry CV5 7FW.
Website: www.cawas.freeserve.co.uk; *Email:* sjp2000@thefarside57.freeserve.co.uk
Meetings: The Earlsdon Church Hall, Albany Road, Earlsdon, Coventry. 2nd Friday, monthly, Sept.–June.

Crawley Astronomical Society
Secretary: Ron Gamer, 1 Pevensey Close, Pound Hill, Crawley, West Sussex RH10 7BL.
Meetings: Ifield Community Centre, Ifield Road, Crawley. 3rd Friday each month, 7.30 p.m.

Crayford Manor House Astronomical Society
Secretary: Roger Pickard, 28 Appletons, Hadlow, Kent TM1 0DT.
Meetings: Manor House Centre, Crayford. Monthly during term time.

Crewkerne and District Astronomical Society (CADAS)
Chairman: Kevin Dodgson, 46 Hermitage Street, Crewkerne, Somerset TA18 8ET.
Email: crewastra@aol.com

Croydon Astronomical Society
Secretary: John Murrell, 17 Dalmeny Road, Carshalton, Surrey.
Meetings: Lecture Theatre, Royal Russell School, Combe Lane, South Croydon. Alternate Fridays, 7.45 p.m.

Derby & District Astronomical Society
Secretary: Ian Bennett, Freers Cottage, Sutton Lane, Etwall.
Web site: www.derby-astro-soc.fsnet/index.html;
Email: bennett.lovatt@btinternet.com
Meetings: Friends Meeting House, Derby. 1st Friday each month, 7.30 p.m.

Doncaster Astronomical Society
Secretary: A. Anson, 15 Cusworth House, St James Street, Doncaster DN1 3AY
Web site: www.donastro.freeserve.co.uk; *Email:* space@donastro.freeserve.co.uk
Meetings: St George's Church House, St George's Church, Church Way, Doncaster. 2nd and 4th Thursday of each month, commencing at 7.30 p.m.

Dumfries Astronomical Society
Secretary: Mr J. Sweeney, 3 Lakeview, Powfoot, Annan DG13 5PG.
Meetings: Gracefield Arts Centre, Edinburgh Road, Dumfries. 3rd Tuesday Aug.–
May, 7.30 p.m.

Dundee Astronomical Society
Secretary: G. Young, 37 Polepark Road, Dundee, Tayside DD1 5QT.
Meetings: Mills Observatory, Balgay Park, Dundee. 1st Friday each month, 7.30 p.m.
Sept.–Apr.

Easington and District Astronomical Society
Secretary: T. Bradley, 52 Jameson Road, Hartlepool, Co. Durham.
Meetings: Easington Comprehensive School, Easington Colliery. Every 3rd Thursday
throughout the year, 7.30 p.m.

East Antrim Astronomical Society
Secretary: Stephen Beasant
Website: www.eaas.co.uk
Meetings: Ballyclare High School, Ballyclare, County Antrim. First Monday each
month.

Eastbourne Astronomical Society
Secretary: Peter Gill, 18 Selwyn House, Selwyn Road, Eastbourne, East Sussex
BN21 2LF.
Meetings: Willingdon Memorial Hall, Church Street, Willingdon. One Saturday per
month, Sept.–July, 7.30 p.m.

East Riding Astronomers
Secretary: Tony Scaife, 15 Beech Road, Elloughton, Brough, North Humberside
HU15 1JX.
Meetings: As arranged.

East Sussex Astronomical Society
Secretary: Marcus Croft, 12 St Mary's Cottages, Ninfield Road, Bexhill-on-Sea, East
Sussex.
Website: www.esas.org.uk
Meetings: St Marys School, Wrestwood Road, Bexhill. 1st Thursday of each month,
8 p.m.

Edinburgh University Astronomical Society
Secretary: c/o Dept. of Astronomy, Royal Observatory, Blackford Hill, Edinburgh.

Ewell Astronomical Society
Secretary: Richard Gledhill, 80 Abinger Avenue, Cheam SM2 7LW.
Website: www.ewell-as.co.uk
Meetings: St Mary's Church Hall, London Road, Ewell. 2nd Friday of each month
except August, 7.45 p.m.

Exeter Astronomical Society
Secretary: Tim Sedgwick, Old Dower House, Half Moon, Newton St Cyres, Exeter,
Devon EX5 5AE.
Meetings: The Meeting Room, Wynards, Magdalen Street, Exeter. 1st Thursday of
month.

Farnham Astronomical Society
Secretary: Laurence Anslow, 'Asterion', 18 Wellington Lane, Farnham, Surrey
GU9 9BA.
Meetings: Central Club, South Street, Farnham. 2nd Thursday each month, 8 p.m.

Foredown Tower Astronomy Group
Secretary: M. Feist, Foredown Tower Camera Obscura, Foredown Road, Portslade, East Sussex BN41 2EW.
Meetings: At the above address, 3rd Tuesday each month. 7 p.m. (winter), 8 p.m. (summer).

Greenock Astronomical Society
Secretary: Carl Hempsey, 49 Brisbane Street, Greenock.
Meetings: Greenock Arts Guild, 3 Campbell Street, Greenock.

Grimsby Astronomical Society
Secretary: R. Williams, 14 Richmond Close, Grimsby, South Humberside.
Meetings: Secretary's home. 2nd Thursday each month, 7.30 p.m.

Guernsey: La Société Guernesiasie Astronomy Section
Secretary: Debby Quertier, Lamorna, Route Charles, St Peter Port, Guernsey GY1 1QS. and Jessica Harris, Keanda, Les Sauvagees, St Sampson's, Guernsey GY2 4XT.
Meetings: Observatory, Rue du Lorier, St Peter's. Tuesdays, 8 p.m.

Guildford Astronomical Society
Secretary: A. Langmaid, 22 West Mount, The Mount, Guildford, Surrey GU2 5HL.
Meetings: Guildford Institute, Ward Street, Guildford. 1st Thursday each month except Aug., 7.30 p.m.

Gwynedd Astronomical Society
Secretary: Mr Ernie Greenwood, 18 Twrcelyn Street, Llanerchymedd, Anglesey LL74 8TL.
Meetings: Dept. of Electronic Engineering, Bangor University. 1st Thursday each month except Aug., 7.30 p.m.

The Hampshire Astronomical Group
Secretary: Geoff Mann, 10 Marie Court, 348 London Road, Waterlooville, Hampshire PO7 7SR.
Website: www.hantsastro.demon.co.uk; *Email:* Geoff.Mann@hazleton97.fsnet.co.uk
Meetings: 2nd Friday, Clanfield Memorial Hall, all other Fridays Clanfield Observatory.

Hanney & District Astronomical Society
Secretary: Bob Church, 47 Upthorpe Drive, Wantage, Oxfordshire OX12 7DG.
Meetings: Last Thursday each month, 8 p.m.

Harrogate Astronomical Society
Secretary: Brian Bonser, 114 Main Street, Little Ouseburn TO5 9TG.
Meetings: National Power HQ, Beckwith Knowle, Harrogate. Last Friday each month.

Hastings and Battle Astronomical Society
Secretary: K.A. Woodcock, 24 Emmanuel Road, Hastings, East Sussex TN34 3LB.
Email: keith@habas.freeserve.co.uk
Meetings: Herstmonceux Science Centre. 2nd Saturday of each month, 7.30 p.m.

Havering Astronomical Society
Secretary: Frances Ridgley, 133 Severn Drive, Upminster, Essex RM14 1PP.
Meetings: Cranham Community Centre, Marlborough Gardens, Upminster, Essex. 3rd Wednesday each month except July and Aug., 7.30 p.m.

Heart of England Astronomical Society
Secretary: John Williams, 100 Stanway Road, Shirley, Solihull B90 3JG.
Website: www.members.aol.com/hoeas/home.html; *Email:* hoeas@aol.com
Meetings: Furnace End Village, over Whitacre, Warwickshire. Last Thursday each
month, except June, July & Aug., 8 p.m.

Hebden Bridge Literary & Scientific Society, Astronomical Section
Secretary: Peter Jackson, 44 Gilstead Lane, Bingley, West Yorkshire BD16 3NP.
Meetings: Hebden Bridge Information Centre. Last Wednesday, Sept.–May.

Herefordshire Astronomical Society
Secretary: Paul Olver, The Buttridge, Wellington Lane, Canon Pyon, Hereford
HR4 8NL.
Email: info@hsastro.org.uk
Meetings: The Kindle Centre, ASDA Supermarket, Hereford. 1st Thursday of every
month (except August) 7 p.m.

Herschel Astronomy Society
Secretary: Kevin Bishop, 106 Holmsdale, Crown Wood, Bracknell, Berkshire
RG12 3TB.
Meetings: Eton College. 2nd Friday each month, 7.30 p.m.

Highlands Astronomical Society
Secretary: Richard Green, 11 Drumossie Avenue, Culcabock, Inverness IV2 3SJ.
Meetings: The Spectrum Centre, Inverness. 1st Tuesday each month, 7.30 p.m.

Hinckley & District Astronomical Society
Secretary: Mr S. Albrighton, 4 Walnut Close, The Bridleways, Hartshill, Nuneaton,
Warwickshire CV10 0XH.
Meetings: Burbage Common Visitors Centre, Hinckley. 1st Tuesday Sept.–May,
7.30 p.m.

Horsham Astronomy Group (was **Forest Astronomical Society**)
Secretary: Dan White, 32 Burns Close, Horsham, West Sussex RH12 5PF.
Email: secretary@horshamastronomy.com
Meetings: 1st Wednesday each month.

Howards Astronomy Club
Secretary: H. Ilett, 22 St George's Avenue, Warblington, Havant, Hampshire.
Meetings: To be notified.

Huddersfield Astronomical and Philosophical Society
Secretary: Lisa B. Jeffries, 58 Beaumont Street, Netherton, Huddersfield, West
Yorkshire HD4 7HE.
Email: l.b.jeffries@hud.ac.uk
Meetings: 4a Railway Street, Huddersfield. Every Wednesday and Friday, 7.30 p.m.

Hull and East Riding Astronomical Society
President: Sharon E. Long
Email: charon@charon.karoo.co.uk
Website: http://www.heras.org.uk
Meetings: The Wilberforce Building, Room S25, University of Hull, Cottingham
Road, Hull. 2nd Monday each month, Sept.–May, 7.30–9.30 p.m.

Ilkeston & District Astronomical Society
Secretary: Mark Thomas, 2 Elm Avenue, Sandiacre, Nottingham NG10 5EJ.
Meetings: The Function Room, Erewash Museum, Anchor Row, Ilkeston. 2nd
Tuesday monthly, 7.30 p.m.

Ipswich, Orwell Astronomical Society
Secretary: R. Gooding, 168 Ashcroft Road, Ipswich.
Meetings: Orwell Park Observatory, Nacton, Ipswich. Wednesdays, 8 p.m.

Irish Astronomical Association
President: Terry Moseley, 31 Sunderland Road, Belfast BT6 9LY, Northern Ireland.
Email: terrymosel@aol.com
Meetings: Ashby Building, Stranmillis Road, Belfast. Alternate Wednesdays,
7.30 p.m.

Irish Astronomical Society
Secretary: James O'Connor, PO Box 2547, Dublin 15, Eire.
Meetings: Ely House, 8 Ely Place, Dublin 2. 1st and 3rd Monday each month.

Isle of Man Astronomical Society
Secretary: James Martin, Ballaterson Farm, Peel, Isle of Man IM5 3AB.
Email: ballaterson@manx.net
Meetings: Isle of Man Observatory, Foxdale. 1st Thursday of each month, 8 p.m.

Isle of Wight Astronomical Society
Secretary: J. W. Feakins, 1 Hilltop Cottages, High Street, Freshwater, Isle of Wight.
Meetings: Unitarian Church Hall, Newport, Isle of Wight. Monthly.

Keele Astronomical Society
Secretary: Natalie Webb, Department of Physics, University of Keele, Keele,
Staffordshire ST5 5BG.
Meetings: As arranged during term time.

Kettering and District Astronomical Society
Asst. Secretary: Steve Williams, 120 Brickhill Road, Wellingborough,
Northamptonshire.
Meetings: Quaker Meeting Hall, Northall Street, Kettering, Northamptonshire.
1st Tuesday each month, 7.45 p.m.

King's Lynn Amateur Astronomical Association
Secretary: P. Twynman, 17 Poplar Avenue, RAF Marham, King's Lynn.
Meetings: As arranged.

Lancaster and Morecambe Astronomical Society
Secretary: Mrs E. Robinson, 4 Bedford Place, Lancaster LA1 4EB.
Email: ehelenerob@btinternet.com
Meetings: Church of the Ascension, Torrisholme. 1st Wednesday each month except
July and Aug.

Knowle Astronomical Society
Secretary: Nigel Foster, 21 Speedwell Drive, Balsall Common, Coventry,
West Midlands CV7 7AU.
Meetings: St George & St Theresa's Parish Centre, 337 Station Road, Dorridge,
Solihull, West Midlands B93 8TZ. 1st Monday of each month (+/– 1 week for Bank
Holidays) except August.

Lancaster University Astronomical Society
Secretary: c/o Students' Union, Alexandra Square, University of Lancaster.
Meetings: As arranged.

Laymans Astronomical Society
Secretary: John Evans, 10 Arkwright Walk, The Meadows, Nottingham.
Meetings: The Popular, Bath Street, Ilkeston, Derbyshire. Monthly.

Leeds Astronomical Society
Secretary: Mark A. Simpson, 37 Roper Avenue, Gledhow, Leeds LS8 1LG.
Meetings: Centenary House, North Street. 2nd Wednesday each month, 7.30 p.m.

Leicester Astronomical Society
Secretary: Dr P.J. Scott, 21 Rembridge Close, Leicester LE3 9AP.
Meetings: Judgemeadow Community College, Marydene Drive, Evington, Leicester.
2nd and 4th Tuesdays each month, 7.30 p.m.

Letchworth and District Astronomical Society
Secretary: Eric Hutton, 14 Folly Close, Hitchin, Hertfordshire.
Meetings: As arranged.

Lewes Amateur Astronomers
Secretary: Christa Sutton, 8 Tower Road, Lancing, West Sussex BN15 9HT.
Meetings: The Bakehouse Studio, Lewes. Last Wednesday each month.

Limerick Astronomy Club
Secretary: Tony O'Hanlon, 26 Ballycannon Heights, Meelick, Co. Clare, Eire.
Meetings: Limerick Senior College, Limerick. Monthly (except June and Aug.), 8 p.m.

Lincoln Astronomical Society
Secretary: David Swaey, 'Everglades', 13 Beaufort Close, Lincoln LN2 4SF.
Meetings: The Lecture Hall, off Westcliffe Street, Lincoln. 1st Tuesday each month.

Liverpool Astronomical Society
Secretary: Mr K. Clark, 31 Sandymount Drive, Wallasey, Merseyside L45 0LJ.
Meetings: Lecture Theatre, Liverpool Museum. 3rd Friday each month, 7 p.m.

Norman Lockyer Observatory Society
Secretary: G.E. White, PO Box 9, Sidmouth EX10 0YQ.
Website: www.ex.ac.uk/nlo/; *Email:* g.e.white@ex.ac.uk
Meetings: Norman Lockyer Observatory, Sidmouth. Fridays and 2nd Monday each month, 7.30 p.m.

Loughton Astronomical Society
Secretary: Charles Munton, 14a Manor Road, Wood Green, London N22 4YJ.
Meetings: 1st Theydon Bois Scout Hall, Loughton Lane, Theydon Bois. Weekly.

Lowestoft and Great Yarmouth Regional Astronomers (LYRA) Society
Secretary: Simon Briggs, 28 Sussex Road, Lowestoft, Suffolk.
Meetings: Community Wing, Kirkley High School, Kirkley Run, Lowestoft. 3rd Thursday each month, 7.30 p.m.

Luton Astronomical Society
Secretary: Mr G. Mitchell, Putteridge Bury, University of Luton, Hitchin Road, Luton.
Website: www.lutonastrosoc.org.uk; *Email:* user998491@aol.com
Meetings: Univ. of Luton, Putteridge Bury (except June, July and August), or Someries Junior School, Wigmore Lane, Luton (July and August only), last Thursday each month, 7.30–9.00 p.m.

Lytham St Annes Astronomical Association
Secretary: K.J. Porter, 141 Blackpool Road, Ansdell, Lytham St Anne's, Lancashire.
Meetings: College of Further Education, Clifton Drive South, Lytham St Anne's. 2nd Wednesday monthly Oct.–June.

Macclesfield Astronomical Society
Secretary: Mr John H. Thomson, 27 Woodbourne Road, Sale, Cheshire M33 3SY
Website: www.maccastro.com; *Email:* jhandlc@yahoo.com
Meetings: Jodrell Bank Science Centre, Goostrey, Cheshire. 1st Tuesday of every month, 7 p.m.

Maidenhead Astronomical Society
Secretary: Tim Haymes, Hill Rise, Knowl Hill Common, Knowl Hill, Reading RG10 9YD.
Meetings: Stubbings Church Hall, near Maidenhead. 1st Friday Sept.–June.

Maidstone Astronomical Society
Secretary: Stephen James, 4 The Cherry Orchard, Haddow, Tonbridge, Kent.
Meetings: Nettlestead Village Hall. 1st Tuesday in the month except July and Aug., 7.30 p.m.

Manchester Astronomical Society
Secretary: Mr Kevin J. Kilburn FRAS, Godlee Observatory, UMIST, Sackville Street, Manchester M60 1QD.
Website: www.u-net.com/ph/mas/; *Email:* kkilburn@globalnet.co.uk
Meetings: At the Godlee Observatory. Thursdays, 7 p.m., except below.
Free Public Lectures: Renold Building UMIST, third Thursday Sept.–Mar., 7.30 p.m.

Mansfield and Sutton Astronomical Society
Secretary: Angus Wright, Sherwood Observatory, Coxmoor Road, Sutton-in-Ashfield, Nottinghamshire NG17 5LF.
Meetings: Sherwood Observatory, Coxmoor Road. Last Tuesday each month, 7.30 p.m.

Mexborough and Swinton Astronomical Society
Secretary: Mark R. Benton, 14 Sandalwood Rise, Swinton, Mexborough, South Yorkshire S64 8PN.
Website: www.msas.org.uk; *Email:* mark@masas.f9.co.uk
Meetings: Swinton WMC. Thursdays, 7.30 p.m.

Mid-Kent Astronomical Society
Secretary: Peter Parish, 30 Wooldeys Road, Rainham, Kent ME8 7NU.
Meetings: Bredhurst Village Hall, Hurstwood Road, Bredhurst, Kent. 2nd and last Fridays each month except August, 7.45 p.m.
Website: www.mkas-site.co.uk

Milton Keynes Astronomical Society
Secretary: Mike Leggett, 19 Matilda Gardens, Shenley Church End, Milton Keynes MK5 6HT.
Website: www.mkas.org.uk; *Email:* mike-pat-leggett@shenley9.fsnet.co.uk
Meetings: Rectory Cottage, Bletchley. Alternate Fridays.

Moray Astronomical Society
Secretary: Richard Pearce, 1 Forsyth Street, Hopeman, Elgin, Moray, Scotland.
Meetings: Village Hall Close, Co. Elgin.

Newbury Amateur Astronomical Society (NAAS)
Secretary: Mrs Monica Balstone, 37 Mount Pleasant, Tadley RG26 4BG.
Meetings: United Reformed Church Hall, Cromwell Place, Newbury. 1st Friday of month, Sept.–June.

Newcastle-on-Tyne Astronomical Society
Secretary: C.E. Willits, 24 Acomb Avenue, Seaton Delaval, Tyne and Wear.
Meetings: Zoology Lecture Theatre, Newcastle University. Monthly.

North Aston Space & Astronomical Club
Secretary: W.R. Chadburn, 14 Oakdale Road, North Aston, Sheffield.
Meetings: To be notified.

Northamptonshire Natural History Society (Astronomy Section)
Secretary: R.A. Marriott, 24 Thirlestane Road, Northampton NN4 8HD.
Email: ram@hamal.demon.co.uk
Meetings: Humfrey Rooms, Castilian Terrace, Northampton. 2nd and last Mondays, most months, 7.30 p.m.

Northants Amateur Astronomers
Secretary: Mervyn Lloyd, 76 Havelock Street, Kettering, Northamptonshire.
Meetings: 1st and 3rd Tuesdays each month, 7.30 p.m.

North Devon Astronomical Society
Secretary: P.G. Vickery, 12 Broad Park Crescent, Ilfracombe, Devon EX34 8DX.
Meetings: Methodist Hall, Rhododendron Avenue, Sticklepath, Barnstaple. 1st Wednesday each month, 7.15 p.m.

North Dorset Astronomical Society
Secretary: J.E.M. Coward, The Pharmacy, Stalbridge, Dorset.
Meetings: Charterhay, Stourton, Caundle, Dorset. 2nd Wednesday each month.

North Downs Astronomical Society
Secretary: Martin Akers, 36 Timber Tops, Lordswood, Chatham, Kent ME5 8XQ.
Meetings: Vigo Village Hall. 3rd Thursday each month. 7.30 p.m.

North-East London Astronomical Society
Secretary: Mr B. Beeston, 38 Abbey Road, Bush Hill Park, Enfield EN1 2QN.
Meetings: Wanstead House, The Green, Wanstead. 3rd Sunday each month (except Aug.), 3 p.m.

North Gwent and District Astronomical Society
Secretary: Jonathan Powell, 14 Lancaster Drive, Gilwern, nr Abergavenny, Monmouthshire NP7 0AA.
Meetings: Gilwern Community Centre. 15th of each month, 7.30 p.m.

North Staffordshire Astronomical Society
Secretary: Duncan Richardson, Halmerend Hall Farm, Halmerend, Stoke-on-Trent, Staffordshire ST7 8AW.
Email: dwr@enterprise.net
Meetings: 21st Hartstill Scout Group HQ, Mount Pleasant, Newcastle-under-Lyme ST5 1DR. 1st Tuesday each month (except July and Aug.), 7–9.30 p.m.

Northumberland Astronomical Society
Contact: Dr Adrian Jametta, 1 Lake Road, Hadston, Morpeth, Northumberland NE65 9TF.
Email: adrian@themoon.co.uk
Website: www.nastro.org.uk
Meetings: Hauxley Nature Reserve (near Amble). Last Thursday of every month (except December), 7.30 pm. Additional meetings and observing sessions listed on website.
Tel: 07984 154904

North Western Association of Variable Star Observers
Secretary: Jeremy Bullivant, 2 Beaminster Road, Heaton Mersey, Stockport, Cheshire.
Meetings: Four annually.

Norwich Astronomical Society
Secretary: Dave Balcombe, 52 Folly Road, Wymondham, Norfolk NR18 0QR.
Website: www.norwich.astronomical.society.org.uk
Meetings: Seething Observatory, Toad Lane, Thwaite St Mary, Norfolk. Every Friday,
7.30 p.m.

Nottingham Astronomical Society
Secretary: C. Brennan, 40 Swindon Close, The Vale, Giltbrook, Nottingham
NG16 2WD.
Meetings: Djanogly City Technology College, Sherwood Rise (B682). 1st and 3rd
Thursdays each month, 7.30 p.m.

Oldham Astronomical Society
Secretary: P.J. Collins, 25 Park Crescent, Chadderton, Oldham.
Meetings: Werneth Park Study Centre, Frederick Street, Oldham. Fortnightly, Friday.

Open University Astronomical Society
Secretary: Dr Andrew Norton, Department of Physics and Astronomy, The Open
University, Walton Hall, Milton Keynes MK7 6AA.
Website: www.physics.open.ac.uk/research/astro/a_club.html
Meetings: Open University, Milton Keynes. 1st Tuesday of every month, 7.30 p.m.

Orpington Astronomical Society
Secretary: Dr Ian Carstairs, 38 Brabourne Rise, Beckenham, Kent BR3 2SG.
Meetings: High Elms Nature Centre, High Elms Country Park, High Elms Road,
Farnborough, Kent. 4th Thursday each month, Sept.–July, 7.30 p.m.

Papworth Astronomy Club
Contact: Keith Tritton, Magpie Cottage, Fox Street, Great Gransden, Sandy,
Bedfordshire SG19 3AA.
Email: kpt2@tutor.open.ac.uk
Meetings: Bradbury Progression Centre, Church Lane, Papworth Everard, nr
Huntingdon. 1st Wednesday each month, 7 p.m.

Peterborough Astronomical Society
Secretary: Sheila Thorpe, 6 Cypress Close, Longthorpe, Peterborough.
Meetings: 1st Thursday every month, 7.30 p.m.

Plymouth Astronomical Society
Secretary: Alan G. Penman, 12 St Maurice View, Plympton, Plymouth, Devon
PL7 1FQ.
Email: oakmount12@aol.com
Meetings: Glynis Kingham Centre, YMCA Annex, Lockyer Street, Plymouth. 2nd
Friday each month, 7.30 p.m.

PONLAF
Secretary: Matthew Hepburn, 6 Court Road, Caterham, Surrey CR3 5RD.
Meetings: Room 5, 6th floor, Tower Block, University of North London. Last Friday
each month during term time, 6.30 p.m.

Port Talbot Astronomical Society (formerly **Astronomical Society of Wales**)
Secretary: Mr J. Hawes, 15 Lodge Drive, Baglan, Port Talbot, West Glamorgan
SA12 8UD.
Meetings: Port Talbot Arts Centre. 1st Tuesday each month, 7.15 p.m.

Portsmouth Astronomical Society
Secretary: G.B. Bryant, 81 Ringwood Road, Southsea.
Meetings: Monday, fortnightly.

Preston & District Astronomical Society
Secretary: P. Sloane, 77 Ribby Road, Wrea Green, Kirkham, Preston, Lancashire.
Meetings: Moor Park (Jeremiah Horrocks) Observatory, Preston. 2nd Wednesday, last Friday each month, 7.30 p.m.

Reading Astronomical Society
Secretary: Mrs Ruth Sumner, 22 Anson Crescent, Shinfield, Reading RG2 8JT.
Meetings: St Peter's Church Hall, Church Road, Earley. 3rd Friday each month, 7 p.m.

Renfrewshire Astronomical Society
Secretary: Ian Martin, 10 Aitken Road, Hamilton, South Lanarkshire ML3 7YA.
Website: www.renfrewshire-as.co.uk; *Email:* RenfrewAS@aol.com
Meetings: Coats Observatory, Oakshaw Street, Paisley. Fridays, 7.30 p.m.

Rower Astronomical Society
Secretary: Mary Kelly, Knockatore, The Rower, Thomastown, Co. Kilkenny, Eire.

St Helens Amateur Astronomical Society
Secretary: Carl Dingsdale, 125 Canberra Avenue, Thatto Heath, St Helens, Merseyside WA9 5RT.
Meetings: As arranged.

Salford Astronomical Society
Secretary: Mrs Kath Redford, 2 Albermarle Road, Swinton, Manchester M27 5ST.
Meetings: The Observatory, Chaseley Road, Salford. Wednesdays.

Salisbury Astronomical Society
Secretary: Mrs R. Collins, 3 Fairview Road, Salisbury, Wiltshire SP1 1JX.
Meetings: Glebe Hall, Winterbourne Earls, Salisbury. 1st Tuesday each month.

Sandbach Astronomical Society
Secretary: Phil Benson, 8 Gawsworth Drive, Sandbach, Cheshire.
Meetings: Sandbach School, as arranged.

Sawtry & District Astronomical Society
Secretary: Brooke Norton, 2 Newton Road, Sawtry, Huntingdon, Cambridgeshire PE17 5UT.
Meetings: Greenfields Cricket Pavilion, Sawtry Fen. Last Friday each month.

Scarborough & District Astronomical Society
Secretary: Mrs S. Anderson, Basin House Farm, Sawdon, Scarborough, North Yorkshire.
Meetings: Scarborough Public Library. Last Saturday each month, 7–9 p.m.

Scottish Astronomers Group
Secretary: Dr Ken Mackay, Hayford House, Cambusbarron, Stirling FK7 9PR.
Meetings: North of Hadrian's Wall, twice yearly.

Sheffield Astronomical Society
Secretary: Darren Swindels, 102 Sheffield Road, Woodhouse, Sheffield, South Yorkshire S13 7EU.
Website: www.sheffieldastro.org.uk; *Email:* info@sheffieldastro.org.uk
Meetings: Twice monthly at Mayfield Environmental Education Centre, David Lane, Fulwood, Sheffield S10, 7.30–10 p.m.

Shetland Astronomical Society
Secretary: Peter Kelly, The Glebe, Fetlar, Shetland ZE2 9DJ.
Email: theglebe@zetnet.co.uk
Meetings: Fetlar, Fridays, Oct.–Mar.

Shropshire Astronomical Society
Contact: Mr David Woodward, 20 Station Road, Condover, Shrewsbury, Shropshire SY5 7BQ.
Website: http://www.shropshire-astro.com; *Email:* jacquidodds@ntlworld.com
Meetings: Quarterly talks at the Gateway Arts and Education Centre, Chester Street, Shrewsbury and monthly observing meetings at Rodington Village Hall.

Sidmouth and District Astronomical Society
Secretary: M. Grant, Salters Meadow, Sidmouth, Devon.
Meetings: Norman Lockyer Observatory, Salcombe Hill. 1st Monday in each month.

Solent Amateur Astronomers
Secretary: Ken Medway, 443 Burgess Road, Swaythling, Southampton SO16 3BL.
Web site: www.delscope.demon.co.uk;
Email: kenmedway@kenmedway.demon.co.uk
Meetings: Room 8, Oaklands Community School, Fairisle Road, Lordshill, Southampton. 3rd Tuesday each month, 7.30 p.m.

Southampton Astronomical Society
Secretary: John Thompson, 4 Heathfield, Hythe, Southampton SO45 5BJ.
Web site: www.home.clara.net/lmhobbs/sas.html;
Email: John.G.Thompson@Tesco.net
Meetings: Conference Room 3, The Civic Centre, Southampton. 2nd Thursday each month (except Aug.), 7.30 p.m.

South Downs Astronomical Society
Secretary: J. Green, 46 Central Avenue, Bognor Regis, West Sussex PO21 5HH.
Website: www.southdowns.org.uk
Meetings: Chichester High School for Boys. 1st Friday in each month (except Aug.).

South-East Essex Astronomical Society
Secretary: C.P. Jones, 29 Buller Road, Laindon, Essex.
Website: www.seeas.dabsol.co.uk/; *Email:* cpj@cix.co.uk
Meetings: Lecture Theatre, Central Library, Victoria Avenue, Southend-on-Sea. Generally 1st Thursday in month, Sept.–May, 7.30 p.m.

South-East Kent Astronomical Society
Secretary: Andrew McCarthy, 25 St Paul's Way, Sandgate, near Folkestone, Kent CT20 3NT.
Meetings: Monthly.

South Lincolnshire Astronomical & Geophysical Society
Secretary: Ian Farley, 12 West Road, Bourne, Lincolnshire PE10 9PS.
Meetings: Adult Education Study Centre, Pinchbeck. 3rd Wednesday each month, 7.30 p.m.

Southport Astronomical Society
Secretary: Patrick Brannon, Willow Cottage, 90 Jacksmere Lane, Scarisbrick, Ormskirk, Lancashire L40 9RS.
Meetings: Monthly Sept.–May, plus observing sessions.

Southport, Ormskirk and District Astronomical Society
Secretary: J.T. Harrison, 92 Cottage Lane, Ormskirk, Lancashire L39 3NJ.
Meetings: Saturday evenings, monthly, as arranged.

South Shields Astronomical Society
Secretary: c/o South Tyneside College, St George's Avenue, South Shields.
Meetings: Marine and Technical College. Each Thursday, 7.30 p.m.

South Somerset Astronomical Society
  Secretary: G. McNelly, 11 Laxton Close, Taunton, Somerset.
  Meetings: Victoria Inn, Skittle Alley, East Reach, Taunton, Somerset. Last Saturday
  each month, 7.30 p.m.

South-West Hertfordshire Astronomical Society
  Secretary: Tom Walsh, 'Finches', Coleshill Lane, Winchmore Hill, Amersham,
  Buckinghamshire HP7 0NP.
  Meetings: Rickmansworth. Last Friday each month, Sept.–May.

Stafford and District Astronomical Society
  Secretary: Miss L. Hodkinson, 6 Elm Walk, Penkridge, Staffordshire ST19 5NL.
  Meetings: Weston Road High School, Stafford. Every 3rd Thursday, Sept.–May,
  7.15 p.m.

Stirling Astronomical Society
  Secretary: Hamish MacPhee, 10 Causewayhead Road, Stirling FK9 5ER.
  Meetings: Smith Museum & Art Gallery, Dumbarton Road, Stirling. 2nd Friday each
  month, 7.30 p.m.

Stoke-on-Trent Astronomical Society
  Secretary: M. Pace, Sundale, Dunnocksfold, Alsager, Stoke-on-Trent.
  Meetings: Cartwright House, Broad Street, Hanley. Monthly.

Stratford-upon-Avon Astronomical Society
  Secretary: Robin Swinbourne, 18 Old Milverton, Leamington Spa, Warwickshire
  CV32 6SA.
  Meetings: Tiddington Home Guard Club. 4th Tuesday each month, 7.30 p.m.

Sunderland Astronomical Society
  Contact: Don Simpson, 78 Stratford Avenue, Grangetown, Sunderland SR2 8RZ.
  Meetings: Friends Meeting House, Roker. 1st, 2nd and 3rd Sundays each month.

Sussex Astronomical Society
  Secretary: Mrs C.G. Sutton, 75 Vale Road, Portslade, Sussex.
  Meetings: English Language Centre, Third Avenue, Hove. Every Wednesday,
  7.30–9.30 p.m., Sept.–May.

Swansea Astronomical Society
  Secretary: Dr Michael Morales, 238 Heol Dulais, Birch Grove, Swansea SA7 9LH.
  Website: www.crysania.co.uk/sas/astro/star
  Meetings: Lecture Room C, Science Tower, University of Swansea. 2nd and 4th
  Thursday each month from Sept.–June, 7 p.m.

Tavistock Astronomical Society
  Secretary: Mrs Ellie Coombes, Rosemount, Under Road, Gunnislake, Cornwall
  PL18 9JL.
  Meetings: Science Laboratory, Kelly College, Tavistock. 1st Wednesday each month,
  7.30 p.m.

Thames Valley Astronomical Group
  Secretary: K.J. Pallet, 82a Tennyson Street, South Lambeth, London SW8 3TH.
  Meetings: As arranged.

Thanet Amateur Astronomical Society
  Secretary: P.F. Jordan, 85 Crescent Road, Ramsgate.
  Meetings: Hilderstone House, Broadstairs, Kent. Monthly.

Torbay Astronomical Society

Secretary: Tim Moffat, 31 Netley Road, Newton Abbot, Devon TQ12 2LL.
Meetings: Torquay Boys' Grammar School, 1st Thursday in month; and Town Hall, Torquay, 3rd Thursday in month, Oct.–May, 7.30 p.m.

Tullamore Astronomical Society

Secretary: Tom Walsh, 25 Harbour Walk, Tullamore, Co. Offaly, Eire.
Website: www.iol.ie/seanmck/tas.htm; *Email:* tcwalsh25@yahoo.co.uk
Meetings: Order of Malta Lecture Hall, Tanyard, Tullamore, Co. Offaly, Eire.
Mondays at 8 p.m., every fortnight.

Tyrone Astronomical Society

Secretary: John Ryan, 105 Coolnafranky Park, Cookstown, Co. Tyrone, Northern Ireland.
Meetings: Contact Secretary.

Usk Astronomical Society

Secretary: Bob Wright, 'Llwyn Celyn', 75 Woodland Road, Croesyceiliog, Cwmbran NP44 2OX.
Meetings: Usk Community Education Centre, Maryport Street, Usk. Every Thursday during school term, 7 p.m.

Vectis Astronomical Society

Secretary: Rosemary Pears, 1 Rockmount Cottages, Undercliff Drive, St Lawrence, Ventnor, Isle of Wight PO38 1XG.
Website: www.wightskies.fsnet.co.uk/main.html;
Email: may@tatemma.freeserve.co.uk
Meetings: Lord Louis Library Meeting Room, Newport. 4th Friday each month except Dec., 7.30 p.m.

Vigo Astronomical Society

Secretary: Robert Wilson, 43 Admers Wood, Vigo Village, Meopham, Kent DA13 0SP.
Meetings: Vigo Village Hall. As arranged.

Walsall Astronomical Society

Secretary: Bob Cleverley, 40 Mayfield Road, Sutton Coldfield B74 3PZ.
Meetings: Freetrade Inn, Wood Lane, Pelsall North Common. Every Thursday.

Wellingborough District Astronomical Society

Secretary: S.M. Williams, 120 Brickhill Road, Wellingborough, Northamptonshire.
Meetings: Gloucester Hall, Church Street, Wellingborough. 2nd Wednesday each month, 7.30 p.m.

Wessex Astronomical Society

Secretary: Leslie Fry, 14 Hanhum Road, Corfe Mullen, Dorset.
Meetings: Allendale Centre, Wimborne, Dorset. 1st Tuesday of each month.

West Cornwall Astronomical Society

Secretary: Dr R. Waddling, The Pines, Pennance Road, Falmouth, Cornwall TR11 4ED.
Meetings: Helston Football Club, 3rd Thursday each month, and St Michall's Hotel, 1st Wednesday each month, 7.30 p.m.

West of London Astronomical Society

Secretary: Duncan Radbourne, 28 Tavistock Road, Edgware, Middlesex HA8 6DA.
Website: www.wocas.org.uk
Meetings: Monthly, alternately in Uxbridge and North Harrow. 2nd Monday in month, except Aug.

West Midlands Astronomical Association
Secretary: Miss S. Bundy, 93 Greenridge Road, Handsworth Wood, Birmingham.
Meetings: Dr Johnson House, Bull Street, Birmingham. As arranged.

West Yorkshire Astronomical Society
Secretary: Pete Lunn, 21 Crawford Drive, Wakefield, West Yorkshire.
Meetings: Rosse Observatory, Carleton Community Centre, Carleton Road, Pontefract. Each Tuesday, 7.15 p.m.

Whitby and District Astronomical Society
Secretary: Rosemary Bowman, The Cottage, Larpool Drive, Whitby, North Yorkshire YO22 4ND.
Meetings: Whitby Mission, Seafarers' Centre, Haggersgate, Whitby. 1st Tuesday of the month, 7.30 p.m.

Whittington Astronomical Society
Secretary: Peter Williamson, The Observatory, Top Street, Whittington, Shropshire.
Meetings: The Observatory. Every month.

Wiltshire Astronomical Society
Secretary: Simon Barnes, 25 Woodcombe, Melksham, Wiltshire SN12 6HA.
Meetings: St Andrew's Church Hall, Church Lane, off Forest Road, Melksham, Wiltshire.

Wolverhampton Astronomical Society
Secretary: Mr M. Bryce, Iona, 16 Yellowhammer Court, Kidderminster, Worcestershire DY10 4RR.
Website: www.wolvas.org.uk; *Email:* michaelbryce@wolvas.org.uk
Meetings: Beckminster Methodist Church Hall, Birches Barn Road, Wolverhampton. Alternate Mondays, Sept.–Apr., extra dates in summer, 7.30 p.m.

Worcester Astronomical Society
Secretary: Mr S. Bateman, 12 Bozward Street, Worcester WR2 5DE.
Meetings: Room 117, Worcester College of Higher Education, Henwick Grove, Worcester. 2nd Thursday each month, 8 p.m.

Worthing Astronomical Society
Contact: G. Boots, 101 Ardingly Drive, Worthing, West Sussex BN12 4TW.
Website: www.worthingastro.freeserve.co.uk;
Email: gboots@observatory99.freeserve.co.uk
Meetings: Heene Church Rooms, Heene Road, Worthing. 1st Wednesday each month (except Aug.), 7.30 p.m.

Wycombe Astronomical Society
Secretary: Mr P. Treherne, 34 Honeysuckle Road, Widmer End, High Wycombe, Buckinghamshire HP15 6BW.
Meetings: Woodrow High House, Amersham. 3rd Wednesday each month, 7.45 p.m.

The York Astronomical Society
Contact: Hazel Collett, Public Relations Officer
Tel: 07944 751277
Website: www.yorkastro.freeserve.co.uk; *Email:* info@yorkastro.co.uk
Meetings: The Knavesmire Room, York Priory Street Centre, Priory Street, York. 1st and 3rd Friday of each month (except Aug.), 8 p.m.

Any society wishing to be included in this list of local societies or to update details, including any website addresses, is invited to write to the Editor (c/o Pan Macmillan, 20 New Wharf Road, London N1 9RR), so that the relevant information may be included in the next edition of the *Yearbook*.

The William Herschel Society maintains the museum established at 19 New King Street, Bath BA1 2BL – the only surviving Herschel House. It also undertakes activities of various kinds. New members would be welcome; those interested are asked to contact the Membership Secretary at the museum.

The South Downs Planetarium (Kingsham Farm, Kingsham Road, Chichester, West Sussex PO19 8RP) is now fully operational. For further information, visit www.southdowns.org.uk/sdpt or telephone (01243) 774400